Micro- and Nanoelectromechanical Biosensors

FOCUS SERIES

Series Editor Pascal Maigné

Micro- and Nanoelectromechanical Biosensors

Liviu Nicu
Thierry Leïchlé

iSTE

WILEY

First published 2014 in Great Britain and the United States by ISTE Ltd and John Wiley & Sons, Inc.

ISTE Ltd
27-37 St George's Road
London SW19 4EU
UK

www.iste.co.uk

John Wiley & Sons, Inc.
111 River Street
Hoboken, NJ 07030
USA

www.wiley.com

Library of Congress Control Number: 2013952645

British Library Cataloguing-in-Publication Data
A CIP record for this book is available from the British Library
ISSN 2051-2481 (Print)
ISSN 2051-249X (Online)
ISBN 978-1-84821-479-8

Printed and bound in Great Britain by CPI Group (UK) Ltd., Croydon, Surrey CR0 4YY

Contents

Introduction

In a world where a biological threat may take multiple forms associated with environmental, health or defense issues, the need for versatile biosensing platforms is of vital concern. The variability of biological matter is proportional to the infinite different ways in which it impacts human beings, in timescales ranging from hours (for particularly aggressive viruses such as those provoking hemorrhagic fevers) to years (for somatic evolution processes leading to cancer). Identification and quantification of one or several biological species of harmful potential are the design targets for the vast majority of the existing biosensors.

No matter what type of biological species (be they viruses, bacteria or circulating proteins in the bloodstream) are targeted by a biosensor, the bottom line of the fundamental requirements for a successful biosensing process is always the same: the best specificity, sensitivity and fastest time of analysis. More specifically, we can add portability, user-friendly exploitation interfaces, cost and a few others which are of secondary concern. In a contemporary technological context where the plethora of configurations seems to meet part or all of the previously listed requirements, it seemed of paramount importance to the authors to reassess the basics of exactly what micromechanics can do in order to overtake the biosensing area, where compatible. This book intends to shed light upon the field of microelectromechanical systems (MEMS)-based biosensors.

I.1. A brief history of biosensors

In his concise but remarkable review of the field of biosensors, Kissinger [KIS 05] looks back to the early days of biosensing (the 1960s and 1970s) to pinpoint that a "sensor seemed to always be a probe of some sort because of systematic association to pH, ion selectivity or oxygen electrodes". Following the old literature, biosensors are found being called bioelectrodes or enzyme electrodes, or biocatalytic membrane electrodes [ARN 88].

More generally, according to the International Union of Pure and Applied Chemistry (IUPAC) recommendations in 1999, "a biosensor is a self-contained integrated receptor-transducer device, which is capable of providing selective quantitative or semi-quantitative analytical information using a biological recognition element".

The critical feature of the biosensor relates to the selectivity for the specific target analyte; this feature directly impacts the specificity or the process of maintaining the selectivity in the presence of other, potential interfering species. The combination of these quality criteria with miniaturization, low cost and essentially real-time measurements in various fields has generated intense commercial interest.

The last 30 years have witnessed an extraordinary growth in research on sensors in general and biosensors in particular. As underlined by Collings and Caruso in their exhaustive review on biosensor advances [COL 97], "an intensively competitive research area is the result of the combined pressure from the traditional well-springs of research and development – science push and market pull". The growth rate of research activities on biosensors is shown in Figure I.1.

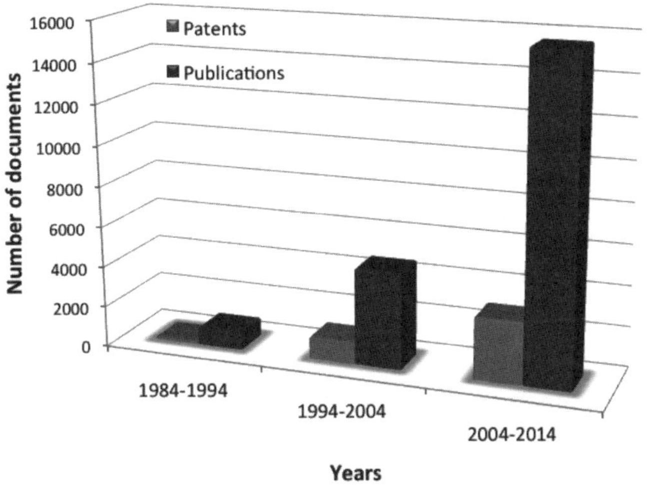

Figure I.1. *Overview of the growth rate of research activities on biosensors since 1984 (Sources: World Intellectual Property Organization database (for patents) and Web of Science Thomson Reuters' database (for publications))*

In spite of all this, there is only one truly commercially successful biosensor: the blood glucose meter for people with diabetes. It is important to note that the glucose biosensor uses the technology that was developed by Clark and Lyons well over 50 years ago and only recently has the public benefited from the potential of such a biosensor. The blood glucose meter is a handheld biosensor based on electrochemical transduction technology [ORA 03] that is produced and commercialized by many companies [TUR 99]. However, in terms of laboratory-based instrumentation, an optical detection system seems to be more commercially viable. Companies such as Affymetrix and Agilent have developed various commercial microarray optical detectors and scanners for genomic and proteomic analysis. Optical sensors that employ surface plasmon resonance (SPR) detection have also been successfully used in many laboratories and universities [RIC 03]. Hence, commercially available optical bench-size immunosensor systems such as BIAcore™ (Biacore AB, Uppsala, Sweden) and IAsys (Affinity Sensors, Cambridge, UK) have found their market in research laboratories for the detection and evaluation of biomolecular interactions.

Still, the development of disposable sensors in conjunction with handheld devices for point of care measurements has featured prominently. Microfabrication technology has played an important part in achieving miniaturized biosensors. Such technology has provided cheap, mass-producible and easy-to-use/disposable sensor strips. Similarly, electrochemical methods have played a pivotal role in detecting the changes that occur during a biorecognition event, and the merging of microfabrication with electrochemical detection has led to the development of various handheld biosensor devices. In fact, i-STAT has developed the world's first handheld device for point-of-care clinical assaying of blood (Figure I.2), noting that this biosensor array employs several electrochemical-based transduction methods (i.e. potentiometric, amperometric and conductometric) [PEJ 06]. However, this is the only example demonstrating the power of microfabrication technologies for the development of biosensors with high integration and multiplex analysis capabilities.

Will markets harvest the fruits of the next generation of biosensors? In fact, it is suggested that a major part of research and development (R&D) activity in this area rarely results in a commercial product [FUJ 04]. However, the observed growth in biosensor research increases the probability of witnessing another success story in the next couple of decades. The future R&D outlook for biosensors looks positive despite very little market growth/progress over the past few years.

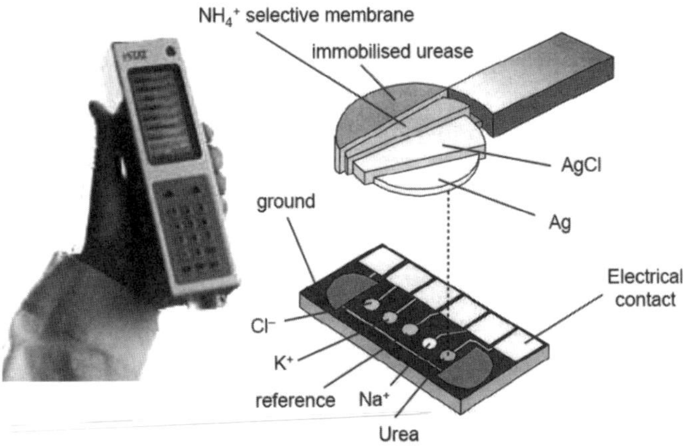

Figure I.2. *The i-STAT multisensor for monitoring various blood electrolytes, gases and metabolites (www.abottpointofcare.com)*

I.2. What is biosensing?

To introduce the field of MEMS biosensors, the concepts and terminology that will be discussed in the next sections and chapters of this book are first to be clarified and described.

I.2.1. *Definitions*

Biosensing: this term is used when a "search and quantify" cycle of operations for one or more biological species (proteins, viruses, bacteria, etc.) is conducted, starting from a sample (either in a gaseous, liquid or solid state) and making use of analytical means of variable complexity.

Biosensor: this is a biosensing device or system made up of two fundamental components: a functionalized solid surface and a transducer which, in turn, transforms a biological reaction (or biological recognition event) taking place on the functionalized surface into a measurable physical signal (Figure I.3).

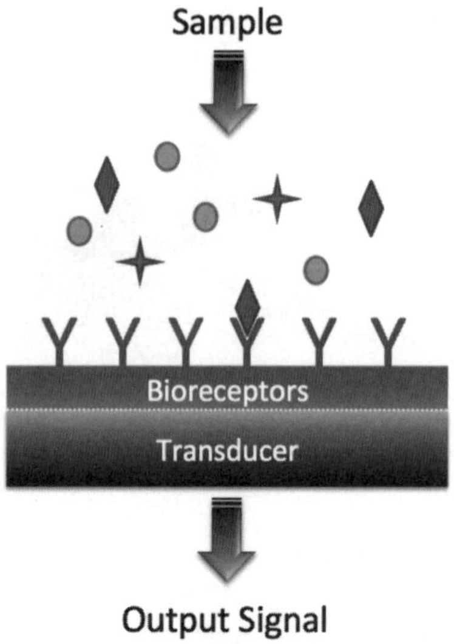

Figure I.3. *Principle and main components of a biosensor*

MEMS biosensor: this is a biosensor using a microelectromechanical system as a transducer.

Multiplexed biosensing: this consists of detecting/dosing several different kinds of biological species present in the same sample, at the same time, in the same fluidic chamber by means of an array of biosensors that are deterministically functionalized prior to the contact with the sample.

Functionalization: this is a succession of chemical and biological reactions on a solid surface, aiming to provide specific reactivity with biological species that are the final targets of a biosensing operation. Functionalization can be either deterministic (i.e. precisely localized) or arbitrary (i.e.

the whole functionalized surface will bear the same chemical and/or biological functionalities).

Fluidics: this is a set of techniques enabling fluid sample circulation toward one or more biosensors. Except for specific requirements, the fluidics operated in most practical cases is "basic fluidics", meaning a reduced volume (several milli-/microliters) closed chamber (or reservoir) made up of biocompatible material (glass, plastic or biologically friendly metals) and bearing access ways (inlets and outlets) which ensure the contact with the outside world by means of flexible capillary tubes.

Sample preparation: this consists of a sequence of steps carried out before sample analysis and aims to render a raw biological sample in a solid, liquid or gaseous phase (e.g. food, blood and air sample) appropriate for biosensing. A clean sample is often required when using sensing techniques that are not responsive to the analyte in its *in situ* form or when the measurement results are distorted by interfering species. Sample preparation includes filtering and separation of unwanted entities (particles, biological entities or chemical species), and dissolution or preconcentration and isolation of the target analyte within an appropriate diluent using various techniques.

I.2.2. *Important numbers and characteristics*

For a biosensor to be a "meet-all-expectations" device, it must fulfill a set of requirements (also known as specifications) that are specific to the biosensing domain. The most important of them are defined as follows:

Sensitivity: this is the transducer output variation induced by the biological interaction events happening on its functionalized surface.

Limit of detectable output (LDO): this indicates the ultimate performance in terms of discernibility of the transducer's output variations with respect to the measurement noise.

Response time: this relates to the time period between the sample injection in the fluidics surrounding the biosensor and the moment when the transducer's output signal is stabilized after having shifted because of the biological interactions effects happening on its functionalized surface.

The list of performances given above is far from being exhaustive. Other specifications exclusively dependent on the format of the transducer might be given. However, we deliberately limit the enumeration to the most generic features that have to be kept in mind and addressed before aiming to use MEMS as biosensors. For an exhaustive classification of biosensors and their associated specifications, readers are referred to [GIZ 02].

I.2.3. *Classification of biosensors*

Biosensors can be classified according to three factors: (1) the type of receptor (e.g. an immunosensor), (2) the physics of the transduction process (e.g. an amperometric sensor) or (3) the application (e.g. a medical biosensor).

The biological recognition element or bioreceptor is a crucial component of the biosensor device. Generally, we can draw three principal classes of biosensors that are distinguished from each other by the nature of the biological process and by the involved biochemical or biological components, i.e. biocatalytic (e.g. enzyme), immunological (e.g. antibody) and nucleic acid (e.g. deoxyribonucleic acid – DNA).

The transducer is another component of the biosensor, which plays an important role in the detection process. A wide variety of transducer methods have been developed in the past decade; however, recent literature reviews have shown that the most popular and common methods presently available are (1) electrochemical, (2) optical, (3) piezoelectric, and (4) thermal or calorimetric [HAL 90, BUE 93, GIZ 02].

These groups can be further divided into general categories: non-labeled or label-free types, which are based on the direct measurement of a phenomenon occurring during the biochemical reactions on a transducer surface, and labeled, which relies on the detection of a specific label that is linked to the analyte to detect. Research into label-free biosensors continues to grow [COO 03]. However, biosensors based on the detection of labeled analytes are more common and are extremely successful in a number of platforms (especially those based on the use of fluorescent labels detected by optical means).

From an application point of view, even if medical and clinical applications are the most lucrative and important avenues for biosensors, other areas like environment, industrial process monitoring and control, or defense require specific biosensing systems. Moreover, commercial biosensors can be divided into two categories on the basis of whether they are laboratory or portable/field devices.

I.3. Biosensing applications and examples

To better assess the impact that MEMS might have on the biosensing domain in the next few years, let us use three fictive examples of potential applications, respectively, in the environmental, health and defense realms. The enumeration of requirements related to real-world situations will allow us to emphasize the advantages of such devices with respect to alternative, already existing solutions. Each of the chosen examples will be presented following the same synoptic basis: (1) context of the raised biological issue and (2) bioMEMS-based optional solution to avoid the bio-crisis. A general conclusion will highlight the added value brought by the bioMEMS solution.

I.3.1. *Environment*

I.3.1.1. *Context of the raised biological issue*

It had been several days since the local university canoeing team set up its training camp on the border of the Blue River, the same as every year, in early summer time, before starting the competition season. This year, the training camp had gathered together the whole team, meaning there were 24 athletes, and 3 trainers. Despite there having been good weather conditions since the beginning of the stay, by the middle of the second week three of the young athletes had started to show signs of evident illness such as fever, chills, myalgia and intense headaches. It mostly looked like a flu-related syndrome, which was all but probable at this time of year.

The decision to break camp and return home was rapidly taken to allow immediate identification of the origins of the illness and to seek treatment as soon as possible. Finally, it took 8 more days (and 6 more infected students) to isolate the cause: leptospirosis, a disease caused by infection with the Leptospira bacteria which is transmitted from animals to people when water contaminated by animal urine comes into contact with the eyes or with the mucous membrane.

I.3.1.2. *How could all this have been avoided?*

Let us imagine the same canoeing team setting up the camp on the Blue River border. Each and every morning, one hour before starting the training day, the same ritual is taking place on the river banks: one of the trainers is pouring few milliliters of water over a small, portable, handheld device. The liquid is driven through a network of flexible capillary tubes while being filtered such that it becomes an inorganic debris-free sample, ready to be analyzed by an array of several dozens of tiny mechanical sensors functionalized with antibodies specific to the most common biological pathogens living in dirty, infected water.

The analysis result is provided within a couple of minutes: if it is positive to one or several pathogens, the measurement protocol is repeated once to avoid a false positive conclusion and if confirmed, the camp is eventually evacuated, thus avoiding potentially infectious exposure of the team and subsequent illness.

I.3.2. *Health*

I.3.2.1. *Context of the raised biological issue*

The same as every day since his wife had been gone, Mr. Jones, 92 years old, had started that morning by making a pot of coffee for breakfast. He was one among the lucky elderly whose autonomy was preserved by a satisfactory overall physical condition providing regular, basic analysis to monitor his health condition. But that morning, unlike the previous ones, Mr. Jones felt like something had gone wrong: a sort of squeeze-like sensation located at the center of his chest was bothering him. Trying not to pay much attention to it, Mr. Jones sat for a few moments on a seat in his kitchen hoping that the chest pain was about to vanish. Instead, his vision became blurred and sweat started to pearl on his forehead. What if the warning was more than serious? He finally grabbed his phone and dialed the emergency number trying to keep calm so he could repeat his symptoms to the operator in a clear manner. Suspecting a heart attack, a medical team was immediately directed to Mr. Jones's address, and half an hour later the primary care protocol was applied. While ensuring that Mr. Jones's airway was normal and it was not laborious for him to breathe, the first results delivered by a handheld electrocardiograph showed that the heart was functioning well. The severe heart attack hypothesis had for the moment been swept away even if cardiac biomarker monitoring had to be done over several hours at the hospital.

Realizing that his heart was finally performing well, Mr. Jones started to feel better and progressively told the medical team about the nightmares he had the night before: he had dreamt of his wife fighting hard against the cancer a few years before and crying for help while he could not do anything. The medical team members started to understand the origins of Mr. Jones's panic and while they kept reassuring him that everything was under control, they decided to transfer him to the hospital for supplementary tests in order to definitely avoid any doubt.

I.3.2.2. *How could all this have been avoided?*

Imagine Mr. Jones at the very beginning of his symptoms of pain and panic grabbing a small suitcase while having a seat in his kitchen. The suitcase had been given to him by his family doctor a few months before and he had been trained to use it in the case of a cardiac alert. Two main elements were packed inside the suitcase: a small, flexible patch to be applied onto the chest and a smartphone-like device having been calibrated to identify and measure cardiac biomarkers (specific to cardiac disorders) in a small quantity of a blood sample. The instructions for use were simple: first, apply the flexible patch to the chest and push an ON/OFF switch, and second, prick one finger and transfer one drop of blood onto a red-colored area of the smartphone-like device, then wait.

While the flexible patch was transmitting, in real time, the electrocardiogram signals to a remote emergency medical team, the main cardiac biomarkers were analyzed by means of an array of specifically functionalized bioMEMS sensors embedded in the smartphone-like device and the results progressively delivered to the same emergency unit. This first emergency assessment would eventually help the remote medical team in the decision-making process, i.e. to send a doctor to perform further investigations rather than sending a whole medical team in a situation that does not necessarily require such an intervention.

I.3.3. *Defense alert*

I.3.3.1. *Context of the raised biological issue*

The Airbus 380, flight AF066, carrying more than 500 passengers from Paris to Los Angeles has been waiting for landing approval since 1 pm and the risk of the carrier running out of kerosene is greatly increasing minute after minute. The reason for the delay is critical; a group of three people took control of the plane several hours before the scheduled landing time and since then they have been seriously threatening to spread unknown, airborne biological agents inside the cabin if their demands are not met immediately. They have requested the liberation of one of their associates who will board the airplane, and after refueling they have demanded permission to take off for a new destination. To prove their resolve, the three men have isolated several passengers on the upper deck and have exposed them to an aerosolized product. Shortly after exposure, almost all of them presented disturbing symptoms like violent nausea and headaches.

A decision was rapidly taken to follow the terrorists' demands and their requests were completely fulfilled. Three days later, the story ended in the liberation of all the hostages on the tarmac of an eastern African airport when it rapidly transpired that the "bio-aggression" had been completely fake. In reality, the three men had used a vomiting aerosol based on chloropicrin which helped creating the illusion of symptoms related to particularly aggressive biological agents spread among the chosen passengers.

I.3.3.2. *How could all this have been avoided?*

Let us imagine the plane's interior walls were embedded with tiny, invisible sensors functionalized such that when put in contact with aerosols bearing the most well-known pathogen species of bacteria, viruses or toxins, a silent alert is emitted and simultaneously transmitted to the pilots and to the destination control tower concerned. In the previous scenario, such sensors would have been silent as no real threat would have been detected, therefore one could have taken the necessary measures to reject the terrorists' demands while rescuing the passengers.

I.3.4. *Highlight of the added-value brought by the MEMS-based biosensing solution*

We have found from the previous examples that there are common situations where groups of people are gathered in remote regions or confined environments, potentially in contact with infectious biological agents and far from any biological analysis facility. In such cases, having biosensing systems to hand that are low cost, portable, intercommunicating, reliable and user-friendly can be of great help in the prevention of long-term exposure of human beings to infectious pathogens. Despite the already existing plethora of biosensors either commercially or in advanced stages of development, there is still an urgent need for biosensing techniques that also allow different situations to be addressed: we attempt to show in this book that new classes of MEMS biosensor satisfy most of such performance requirements.

I.4. Organization of the book

In fact, a MEMS biosensor consists of three components. As previously discussed, the first two fundamental components are *the functional layer*, which recognizes the biological stimulus, and the *transducer*, which converts the stimulus into a useful, measurable output. But a biosensor would be useless without the *system packaging* (third component) that ensures the adequate operation of the biosensor and its connection to the outside world (e.g. with the end user and the sample to analyze). The latter includes, if necessary, but not exhaustively, a power source, the associated electronics, signal processors or optical components, any type of display or user interface, the associated fluidics for sample preparation and delivery to the bioreceptor layer, and the shell of the device.

We have deliberately chosen to focus this book on the core of a biosensor, i.e. the functional layer and the transduction method. The packaging aspects are not addressed here, even if issues related to the packaging of MEMS biosensors are of paramount importance when considering prototyping and bringing a new product to the market. The first reason regarding this choice comes from the fact that most of these issues are application or product-dependent and the second reason is that one could write books concerning each of the packaging concerns that are far beyond the scope of this book.

Chapter 1 presents the main transducing schemes used in biosensing and amendable to miniaturization with a specific emphasis on mechanical transducers, the foundations of MEMS biosensors. Bioreceptors and means to tether them onto a surface are the subject of Chapter 2. Chapter 3 provides a description of the technological tools available to achieve the specific and localized biofunctionalization of the sensor surface. Chapter 4 discusses the amazing advantages offered by further device miniaturization from MEMS to nanoelectromechanical (NEMS) biosensors in terms of performances and integration capabilities, along with the associated challenges inherent to, e.g., the realization and functionalization of nanoscale structures. Finally, we conclude this book by raising an important issue that must be faced once we have several biosensing technologies, especially miniaturized ones, in hand: how to compare performances of biosensing platforms, and how to choose the most appropriate one for a given application.

I.5. Bibliography

[ARN 88] ARNOLD M.A., MEYERHOFF M.E., "Recent advances in the development and analytical applications of biosensing probes", *Critical Review in Analytical Chemistry*, vol. 20, pp. 149–196, 1988.

[BUE 93] BUERK D.G., *Biosensors: Theory and Applications*, Technomic Publishing Company, 1993.

[COL 97] COLLINGS A.F., CARUSO F., "Biosensors: recent advances", *Reports on Progress in Physics*, vol. 60, pp. 1397–1445, 1997.

[COO 03] COOPER M.A., "Label-free screening of bio-molecular interactions", *Analytical and Bioanalytical Chemistry*, vol. 377, pp. 834–842, 2003.

[DOR 03] D'ORAZIO P., "Biosensors in clinical chemistry", *Clinica Chimica Acta*, vol. 334, pp. 41–69, 2003.

[FUJ 04] FUJI-KEIZAI USA, Inc., U.S. & Worldwide: Biosensor market, R&D, applications and commercial implication, New York, 2004.

[GIZ 02] GIZELI E., LOWE C.R, *Biomolecular Sensors*, Taylor & Francis, 2002.

[HAL 90] HALL E.A.H., *Biosensors*, Open University Press, 1990.

[KIS 05] KISSINGER P.T., "Biosensors – a perspective", *Biosensors Bioelectronics*, vol. 20, pp. 2512–2516, 2005.

[PEJ 06] PEJCIC B., De MARCO R., PARKINSON G., "The role of biosensors in the detection of emerging infectious diseases", *Analyst*, vol. 131, pp. 1079–1090, 2006.

[RIC 03] RICH R.L., MYSZKA D.G., "A survey of the year 2002 commercial optical biosensor literature", *Journal of Molecular Recognition*, vol. 16, pp. 351–382, 2003.

[TUR 99] TURNER A.P.F., CHEN B., PILETSKY S.A., "In vitro diagnostics in diabetes: meeting the challenges", *Clinical Chemistry*, vol. 45, pp. 1596–1601, 1999.

1

Transduction Techniques for Miniaturized Biosensors

Biosensors can be classified on the basis of the transduction methods they use. Transduction may be accomplished via a great variety of methods. Since most forms of biosensing-related transductions can be categorized into one of the following three classes: (1) optical detection, (2) electro(chemical) detection and (3) mechanical detection, the discussion that follows will be articulated around these key methods with a strong emphasis on the last one as it is the main transduction scheme of microelectromechanical system (MEMS) biosensors. Because each of these three classes contains many different subclasses resulting in a large number of possible combinations, only the main detection methods within each class will be highlighted and the progressive miniaturization of the biosensing platforms will be addressed. Miniaturized MEMS biosensors being the main topic of this book, this chapter will start with a brief introduction of the field of biological microelectromechanical systems (bioMEMS) before presenting the various classes of transducers.

1.1. Definition of bioMEMS

BioMEMS can be defined as electromechanical devices or systems fabricated using micro-/nanoscale technologies and dedicated to the manipulation, the analysis or the assembly of biological and chemical entities. Areas of research and applications of bioMEMS range from diagnostics, microfluidics, tissue engineering, surface modification, drug delivery, implantable systems, etc., as witnessed by the number of review articles dealing with MEMS for biology and medicine [KOV 98, POL 00, BAS 04, BAS 06]. The term MEMS is now used so broadly that devices that do not have any electromechanical components (such as DNA and protein arrays) are also sometimes improperly called bioMEMS.

Three classes of materials are generally used to fabricate bioMEMS: (1) microelectronics-related materials such as silicon, glass, metals, etc., (2) plastic and polymeric materials such as polydimethylsiloxane (PDMS) or epoxy-based photoresists (SU-8) and (3) biological materials and entities such as tissues, cells and proteins. The first class of materials has been extensively reported, both from research and industrial points of view, and has traditionally been used for the fabrication of MEMS devices [POL 00, BAS 04, BAS 06]. The second class of materials has been exclusively promoted by microfluidics applications and was shown to be very attractive due to an increased biocompatibility, ease of prototyping [XIA 98], low cost and ability to integrate functional hydrogel materials [PEP 86]. The work encompassing the third class of materials remains relatively unexplored, although it represents new and exciting possibilities at the frontier between bioMEMS and bionanotechnology such as the application of micro- and nanotechnology for tissue engineering [BHA 99].

In the following sections, presenting the main transduction methods used in the field of biosensing, a few examples of miniaturized sensors and bioMEMS, generally based on the use of the first class of materials (microelectronics related), will be discussed.

1.2. Transduction techniques

1.2.1. *Optical transduction*

Optical transducers offer the largest number of possible subcategories of all three transducer classes presented here [HAL 90, BUE 93, GIZ 02]. This is due to one of the major advantages of optical sensors, which is their ability to probe surfaces and films in a non-destructive manner. Moreover, they offer advantages in speed, safety, sensitivity and robustness, as well as permitting *in situ* sensing and real-time measurements. The various types of optical transducers exploit properties such as light absorption, fluorescence / phosphorescence, bio- / chemoluminescence, reflectance, Raman scattering and index of refraction.

It is completely beyond the scope of this book to provide an exhaustive review of the optical transducers and because of the importance of this technology for immunoassays, we have decided to specifically and solely present the surface plasmon resonance (SPR). Briefly, SPR is an optical reflectance procedure that is sensitive to changes in the optical properties of the medium in the vicinity of a metal surface [OTO 68, KRE 68]. A surface plasmon is an electromagnetic field charge density oscillation that can exist at a metal–dielectric interface. Upon excitation of the surface plasmons, an electromagnetic field is formed, which decays exponentially with the distance from the metal film surface into the interfacing medium. When resonance occurs, the reflected light intensity from the metal surface goes through a minimum at a defined angle of incidence, known as the plasmon resonance angle.

Since 1977 when Pockrand *et al.* [POC 77] were one of the first groups to apply SPR as a sensing technique to thin organic assemblies of cadmium arachidate deposited onto silver surfaces, this technique has been widely used for sensing applications either in gas [NYL 82] or liquid media. Despite numerous studies using SPR to examine the interactions between antigens and antibodies for biosensing applications, only more than 20 years after the first description of the phenomenon has a real-time immunosensor become commercially available (BIAcore™, Pharmacia Biosensor AB, Uppsala, Sweden). The biosensor group from Uppsala in Sweden has described the use of SPR for the detection of various proteins on a sensor chip consisting of a thin gold film deposited onto a glass prism and covered with a carboxymethyl-dextran hydrogel matrix [LÖF 90, JOH 91, LÖF 91]. The hydrogel matrix provides covalent linkage sites for antibody binding and also protects the metal film from non-specific adsorption. In addition, this polymer matrix increases the amount of antibody loading compared to that of a flat surface. The BIAcore system uses a flow injection cell, which permits kinetic measurements over a wide dynamic range, and significantly reduces the incubation time required for interaction analyses compared with static solid–liquid interface systems.

The sensitivity of SPR-based biosensors can be very high [LIE 83] with nanomolar concentrations of proteins of molecular weight larger than 10^4 Da being detected. The sensitivity of the SPR system depends on the molecular weight of the analyte since it is the physical amount of bound material that generates the signal. Thus, the SPR technique has the advantage of not requiring labeled molecules. Besides, the real-time measurement enables us to estimate the affinity of two ligands through the determination of binding constants. The major drawback with SPR sensing at present is that the commercially available instrumentation is extremely expensive, and since the sensitivity depends on the optical thickness of the adsorbed layer, small molecules cannot be measured at low concentrations.

Maybe the best way to find a transition toward miniaturized optical biosensors at this point is through the portability of the SPR instruments. For instance, one of the most recent commercialized BIAcore instruments (BIAcore X100) weighs approximately 47 kg, it is thus difficult to imagine that the same system could be easily moved from one point-of-care to another. Although attempts have been made to miniaturize SPR instruments (thus rendering them portable [CHI 07] and even potentially drone carried

[NAI 05], as illustrated in Figure 1.1), the integration levels appropriate to miniaturized optical biosensors are still difficult to achieve.

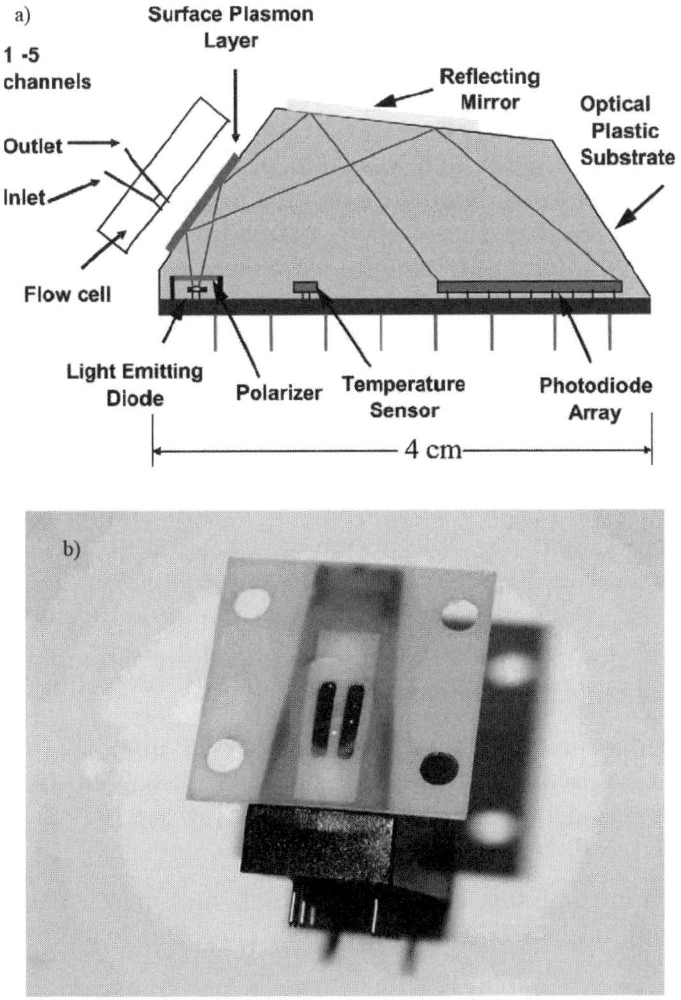

Figure 1.1. *a) Diagram of a miniaturized SPR sensor module. b) Mylar mask with two wells used to functionalize the individual channels of the SPR sensors [NAI 05]*

One representative example of miniaturized optical biosensors is the biosensor developed by the recent joint work of Centro Nacional de Microelectronica and Ikerlan institutes in Spain [SEP 06]. This optical biosensor is based on integrated Mach–Zehnder interferometers, which have been designed to have high surface sensitivity and monomode behavior. As a biosensing application of the devices, the real-time detection of the covalent immobilization and hybridization of DNA strands without labeling has been demonstrated. To achieve a lab-on-a-chip portable microsystem, the integration of the sensor with a complementary metal-oxide-semiconductor (CMOS) compatible microfluidic system using SU-8 photolithography patterned layers was successfully proven.

1.2.2. *Electro (chemical) transduction*

Bioassays with electrochemical detection have gained growing attention since the 1990s [GRE 91, HO 95, NIW 95, BIE 96] because it is possible to measure currents with extremely high precision even in colored and turbid samples (which is practically not achievable using optical transduction). There are the following three main classes of electrochemical biosensors:

1) Amperometric biosensors, which involve the electric current associated with the electrons involved in redox processes (fairly sensitive, currents as low as 10^{-10} A can be recorded with commercial devices).

2) Potentiometric biosensors, which measure a change in potential at electrodes due to ions or chemical reactions at an electrode.

3) Conductometric biosensors, which measure conductance changes associated with changes in the overall ionic medium between two electrodes.

Under optimum conditions – with high enzyme loading under fast mass transport in thin layers and efficient external mass transfer – an enzyme electrode can measure analytes down to 10 µM with acceptable precision. To further increase the sensitivity (down to the nanomolar range), the enzymatic substrate regeneration technique makes use of continuous regeneration of the analyte in cyclic reactions.

A potentiometric biosensor for protein and amino acid estimation was reported by Sarkar and Turner [SAR 99]: a screen-printed biosensor based on a rhodinized carbon-paste working electrode that was used in a three-electrode configuration (reference, working electrode and counter electrode) for a two-step detection method. Electrolysis of an acidic potassium bromide electrolyte at the working electrode produced bromine that was consumed by the proteins and amino acids involved in the assay. The bromine production occurred at one potential while the monitoring of the bromine consumption was performed using a lower potential. The method proved to be very sensitive to almost all of the amino acids, as well as some common proteins in fruit juice, milk and urine, consuming approximately 10 µl of sample for direct detection.

An electrochemical flow-through enzyme-based biosensor for the detection of glucose and lactate has also been developed by Rudel *et al.* [RUD 96]. In this application, glucose oxidase and lactase oxidase were immobilized in conducting polymers generated from pyrrole, *N*-methylpyrrole, aniline and o-phenylenediamine on platinum surfaces. Various sensor matrices were compared based on amperometric measurements of glucose and lactate and it was found that the o-phenylenediamine was the most sensitive polymer. This polymer matrix was then deposited on a piece of graphite felt and used as an enzyme reactor as well as a working electrode in an electrochemical detection system. Using this system, a linear dynamic range of 500 µM–10 mM glucose was determined with a limit of detection <500 µM. For lactate, the dynamic range covered concentrations from 50 µM to 1 mM with a detection limit <50 µM.

Electrical or electrochemical detection techniques can be amenable to portability (through miniaturization) as witnessed by several potentiometric and conductimetric sensors that have been transposed to the micro- and even the nanoscale. The most common form of potentiometric microsensor is the ion-sensitive field effect transistor (ISFET) or chemical field effect transistor (ChemFET). These devices are available commercially as pH sensors and many examples of their use have been reported in the literature [SCH 96]. For instance, potentiometric sensors with ion-selective ionophores in modified poly(vinyl chloride) have been used to detect analytes from human sera [HIN 95]. Cellular respiration and acidification due to the activity of colon adenocarcinoma cell lines have been measured with CMOS ISFET [LEH 01]. Potentiometric sensors have been downscaled to nanometer dimension through the use of silicon nanowires [CHI 01] (Figure 1.2) and carbon nanotubes [BES 03] as field effect sensors.

Microfabricated conductimetric sensors were also fabricated and used to measure, for instance, extracellular neuronal activity for a long period of time [GRO 77, BOR 97]. Conductance techniques are attractive due to their simplicity and ease of use (since a specialized reference electrode is not needed), and have thus been used to detect a wide variety of entities such as agents of biothreat [MUH 03], biochemicals [SUZ 01] and nucleic acids [DRU 03]. Moreover, conductimetric sensors provide information on the ionic strength of electrolytes and can provide convenient selectivity if coupled with enzyme membranes, as in the case of the detection of glucose [SHU 94] and urea [STE 97], for instance.

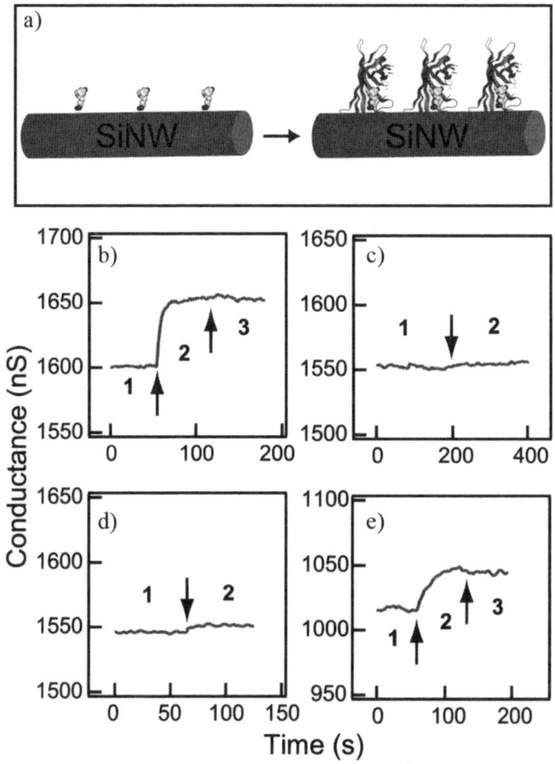

Figure 1.2. *Real-time detection of protein binding with a silicon nanowire (SiNW). a) Schematic illustration of a biotin-modified SiNW (left) and subsequent binding of streptavidin to the SiNW surface (right). b) Plot of conductance versus time for a biotin-modified SiNW, where regions 1, 2, and 3, respectively, correspond to buffer solution, addition of 250 nM streptavidin and pure buffer solution. c) Conductance versus time for an unmodified SiNW; regions 1 and 2 are the same as in (b). d) Response of the biotin-modified sensor to the addition of a 250 nM streptavidin solution that was preincubated with four equivalents d-biotin. e) Response of the biotin-modified sensor to the addition of 25 pM streptavidin [CHI 01]*

A few years ago, more than half of the biosensors reported in the literature were based on electrochemical transducers [MEA 96]. This may not be surprising considering that electrochemical transduction possesses the advantages highlighted in the previously discussed applications: low cost, high sensitivity, independence from solution turbidity, easily miniaturized/well suited to microfabrication, low power requirements and relatively simple instrumentation. These characteristics make electrochemical transduction methods highly compatible for implantable and/or portable hand-held devices.

1.2.3. *Mechanical transduction*

Most of the time, mechanical bio(chemical)sensors are referred to as "mass sensitive" due to gravity or thickness measurements of thin rigid films in the vacuum or gaseous environment for which they were employed at first. Although the mass effect is only one of several contributions to the sensor response in liquids, particularly in the case of soft biological layers, this terminology will be used throughout this section to stress the mechanical basis of the acoustic transduction compared to the other transduction methods previously discussed.

Acoustic wave devices give information on two physical quantities: the acoustic energy storage and the energy dissipation, which can, respectively, be measured by the frequency shift and the damping of the acoustic oscillation. Therefore, more than one signal can be obtained from a single sensor (note that acoustic resonators exhibit infinite resonant modes of vibration, each characterized by a specific resonant frequency and damping factor). This paves the way toward multicomponent signal analysis with a small number of different sensors.

In the last few years, the interest in using acoustic wave devices as biosensors has increased for studying kinetic and thermodynamic parameters of many biochemical or biological systems as well as in the field of understanding the acoustic transduction mechanisms in terms of mechanical interfacial processes. This section is structured as follows: first, physics of piezoelectric excited acoustic waves in solids is briefly discussed, then a general classification of piezoelectric mass-sensing devices is given and finally, we present an example of commercial success based on thickness shear mode resonators before specifically addressing MEMS biosensors in the following section.

1.2.3.1. *Piezoelectric excited acoustic waves in solids*

Piezoelectricity (or the direct piezoelectric effect) as first reported in 1880 by the Curie brothers describes the generation of electrical charges on the surface of a solid caused by pulling, pressure or torsion. In contrast, the occurrence of a mechanical deformation arising from an external field is called the converse piezoelectric effect.

For the piezoelectric effect to potentially occur in a crystalline material, it must not exhibit a center of symmetry. Of the 32 crystal classes, 21 are non-centrosymmetric, and of these, 20 exhibit direct piezoelectricity. Although a large number of crystals show piezoelectricity, only quartz provides the unique combination of mechanical, electrical, chemical, and thermal properties, which has led to its commercial success.

1.2.3.2. *Classification of acoustic piezoelectric resonators*

Usually, piezoelectric devices consist of quartz; other piezoelectric materials are lithium niobate or tantalate, oriented zinc oxide (ZnO), aluminum nitride (AlN) or lead zirconate titanate (PZT). Substrate material, crystal cut (if any) and electrode geometry determine the type of acoustic wave and device. Moreover, acoustic waves are discriminated by the particle displacement direction either relative to the propagation direction of the wave (longitudinal and transverse) or relative to the surface (horizontal and vertical). Bulk waves travel unguided through the volume of the material, surface waves are guided along a device surface and plate waves are guided by reflection from multiple surfaces. In longitudinal waves, particle displacement and wave propagation direction are parallel (compressional waves) whereas in transverse waves they are perpendicular to each other (shear waves). To characterize surface and plate shear waves, we can also distinguish vertical waves with a particle displacement normal to the surface and horizontal waves in which the displacement is parallel to the surface.

Because of the severe damping of acoustic waves in liquid, only devices based on oscillations with particle displacement parallel to the surface can be used for liquid sensing. Surface acoustic wave (SAW) devices use Rayleigh waves with both horizontal and vertical components and are therefore not able to be driven in liquids. An exception from the above stated rule are the flexural plate wave (FPW) devices because the velocity of the antisymmetric Lamb wave is too small for coupling to compressional waves in liquids [JAN 00].

All surface and plate wave devices mostly use interdigitated electrodes (IDE) as transmitters and receivers of the acoustic waves. In contrast, the most frequently used device, the transverse shear mode resonator (TSMR), often called quartz crystal microbalance (QCM), consists of a quartz plate with metal electrodes on each side, which generate bulk waves traveling perpendicular to the sensor surface. A typical recently successful commercialized QCM setup is described in the following.

1.2.3.3. *The quartz crystal microbalance with dissipation monitoring (QCM-D)*

The use of piezoelectric transducers in biosensors was foreshadowed in the work of Sauerbrey [SAU 59] who not only pioneered the use of the QCM, but also thoroughly analyzed the physics of the device. Thus, he demonstrated that there is proportionality between mass added to the QCM sensor and the shift in its resonant frequency.

The major advantage of the QCM sensor derives from the extremely high precision with which the resonance frequency can be measured. Under ideal conditions (temperature and pressure control), the mass sensitivity is down to tens of picograms per square centimeter. The mass sensitivity can be expressed as the so-called Sauerbrey equation:

$$\Delta m = \rho_f \delta_f = -\frac{t_q \rho_q}{f} \Delta f = \frac{C}{n} \Delta f \qquad [1.1]$$

where ρ_f and δ_f are the density and the thickness of the added film, ρ_q and t_q are the density and thickness of the quartz plate, respectively, f is the operation frequency of the sensor, n is the overtone number and C is the so-called mass sensitivity (depending on the rate propagation of the elastic transverse wave in, and the density of, quartz) of the QCM.

The first application of the QCM was as a film thickness monitor for film deposition in vacuum systems or as an oxidation rate sensor [KAS 78, KAS 44, KIN 64]. Other early applications targeted gas sensing in air [GUI 81]. In 1985, Kanazawa opened the door to a whole new set of applications by demonstrating that the QCM could be operated in a stable and reproducible manner in a liquid [KAN 85]. The, by far, most important application of this liquid phase QCM has, until the late 1990s, been the electrochemical QCM (EQCM) [TAK 97].

In spite of many applications of the QCM, it also became successively more evident that a single parameter measurement, i.e. the frequency measurement only, had some severe limitations, especially when non-rigid films were studied. This last point rapidly led to a reconsideration of the Sauerbrey model. Indeed, a condition for the Sauerbrey model to be valid is that the added film perfectly couples to the shear oscillation of the sensor (as, for example, in case of metals or ceramic films deposited onto the active surface of the QCM). However, when the QCM started to be used for studying soft films (as biofilms), the Sauerbrey relation often failed and led to confusion and misinterpretation of data.

A qualitative interpretation of why the Sauerbrey relation loses applicability when the applied films are soft can be seen as follows [GIZ 02]: under such circumstances, the film does not follow the sensor's mechanical oscillation as a "dead mass", but is deformed in the shear direction, such as "a jelly lying on a plate". The frequency shift caused by the jelly-like soft film does not correspond to its static mass but to a dynamic value that depends in a complex manner on its viscous and elastic components. This is the case for soft polymer or biomolecules films that can therefore not be treated as rigid films obeying the Sauerbrey equation, thus implying that the energy dissipation of the QCM may be affected.

Höök [HOO 97] has demonstrated how the simultaneous measurement of both the resonant frequency, *f*, and the dissipation, *D* (which is the reverse of the well-known quality factor, *Q*), helps in the interpretation of the frequency shift in terms of mass uptake and, for the first time, he showed that information about the conformational state of adsorbed protein molecules onto a specifically functionalized QCM sensor could be achieved (Figure 1.3).

This work, in turn, opened up the possibility to extract information about the viscous and elastic components of a soft film added onto a QCM sensor, thus paving the way toward a new technique (called QCM-D, where D stands for dissipation). Systems using the QCM-D technique are now successfully commercialized by Q-Sense (www.q-sense.com, Gothenburg, Sweden).

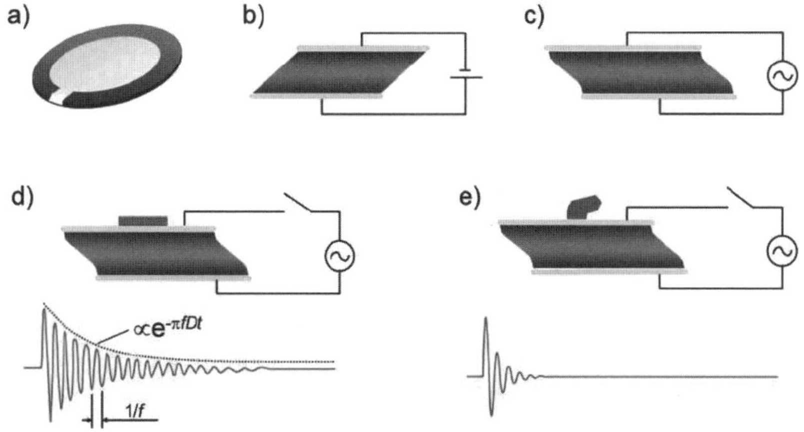

Figure 1.3. *Schematic presentation of the QCM-D working principle. a) The piezoelectric quartz crystal is sandwiched between two gold electrodes. b) The application of an electric field across the crystal results in shear motion of the crystal. c) Resonance in the shear motion can be excited with an oscillating field of appropriate frequency. d) and e) After cutting the driving circuit, the freely decaying oscillation of the crystal is monitored. The temporal change in the crystal's movement, A(t), can be fitted with* $A(t) = A \times exp(-\pi f D t) \times sin(2\pi f t + \varphi)$ *in order to extract the resonance frequency, f, and the dissipation, D. Attachment of a rigid mass d) to the crystal's surface will only lead to a decrease in f, while a soft (viscoelastic) mass e) will also affect D. Monitoring changes in f and D thus enables to follow interfacial processes in real time [RIC 04]*

The adsorption of hemoglobin (Hb) was chosen as a first model in order to illustrate the performance of the QCM-D principle [HOO 98a, HOO 98b], where both Δf and ΔD are simultaneously recorded as a function of time. The resonant frequency and the dissipation factor variations are shown in Figure 1.4 for the adsorption of Hb at a concentration of 1.3 µM on a hydrophobic, methyl terminated thiol-covered gold surface in a 10 mM Hepes solution at pH 6.5 and 7.5 [HOO 98a].

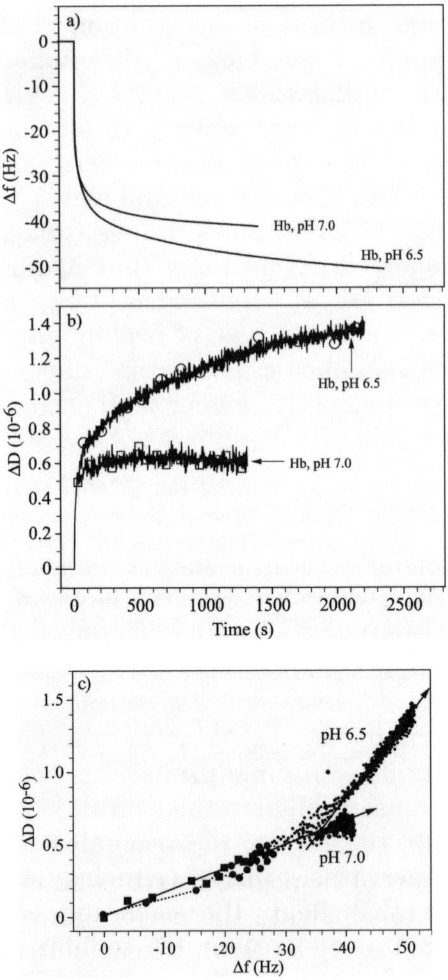

Figure 1.4. *a) and b) Δf versus t and ΔD versus t for the adsorption of Hb on an hydrophobic surface at pH 6.5 and 7.0. c) The influence of pH on the Hb adsorption kinetics is also illustrated by the D-f plot. The simple linear behavior at pH 7.0 is replaced by a two-phase behavior at pH 6.5, where the total mass uptake and final dissipation shift are larger.* (●) *pH 7.0 and* (■) *pH 6.5 [HOO 98a]*

It can be noted that the adsorption causes an initial rapid frequency decrease (equivalent to a mass uptake) at both pH values, followed by a slower frequency decrease as the surface coverage saturates. In addition, the *D*-shifts are positive (meaning an increase of dissipative effects) and display kinetics similar to the *f*-shifts. A useful illustrative way to display the *D*- and *f*-data is to plot Δf versus ΔD to eliminate time as an explicit parameter. It is interesting to note in Figure 1.4(c) that the fast and slow phases of the Hb adsorption cause different slope values (relative dissipation per unit frequency shift) meaning that Δf and ΔD measure different properties of the adsorption kinetics. This illustrates that one single type of protein can form adlayers with different viscoelastic properties depending on their interaction with the surface and/or with each other.

1.3. MEMS transducers

Although friendly to use and ruggedized, the QCM-D technique still suffers from its lack of multiplexing capabilities (recently taken into account in the E4 system configuration from Q-Sense) and most importantly, its lack of integrability. Moreover, improvement of sensitivity remains the driving force in biosensor development. Since the mass sensitivity of resonant sensors is basically governed by the ratio of mass change to the overall vibrating mass, as a rule of thumb, the lighter the resonator is, the more the relative mass increases. Silicon technology can lead to new possibilities in this field: the capability to detect even smaller mass, but also the capability to fabricate arrays with a large number of elements per unit area [ZIM 01, RAB 03], the capability to monolithically integrate electronic circuitry and to mass produce devices at low cost.

Nowadays, MEMS-based mass sensors that consist of commercial cantilevers typically made up of silicon, silicon nitride or silicon dioxide are available. One of the main advantages of cantilever sensors, such as all mass-sensing techniques for biodetection applications, is the ability to detect interacting compounds without the need for introducing an optically detectable label on the analyte.

Historically, exhaustive trials of integrating move-and-sense functions at the microscale have been implemented. Although it is beyond the scope of this chapter to review all actuation/detection of oscillation principles, we cannot ignore the electrostatic/capacitive actuators/sensors widely used as exquisite inertial instruments (accelerometers, gyrometers, pressure sensors) [YAZ 98]. Unfortunately, since biological detection necessarily means operating in a liquid environment, this requirement instantly eliminates capacitive sensors (ideally operated under air or vacuum conditions). Subsequently, other actuating/sensing means have been tested that are more adapted to biosensing: magnetomotive (excellent sensors but integration-limited by the need of an external magnet [VAN 07]), piezoresistive (highly sensitive but ineffective as an actuator [RAS 03]), piezoelectric (excellent actuator but a delicate sensor as it requires high-gain/wide-band charge-amplifying solutions [AYE 08]).

In recent years, very exciting and significant advances in biochemical detection have been made using MEMS. Direct, label-free detection of DNA has been demonstrated using silicon cantilevers through stress-sensing mode (i.e. the real-time monitoring of the cantilevers' bending that translates a change in surface free energy induced by the adsorption of biomolecules on one of its specifically functionalized sides [FRI 00]). These sensors have also been used to sense proteins and cancer markers such as prostate-specific antigen, which have been detected at 0.2 ng/mL in the background of human sera in clinically relevant conditions [WU 01] (Figure 1.5). Recently, challenging experiments have been proposed aiming at monitoring conformational changes of proteins (either bacteriorhodopsin [BRA 06] or human estrogen receptors [MUK 05]) fixed onto microcantilevers.

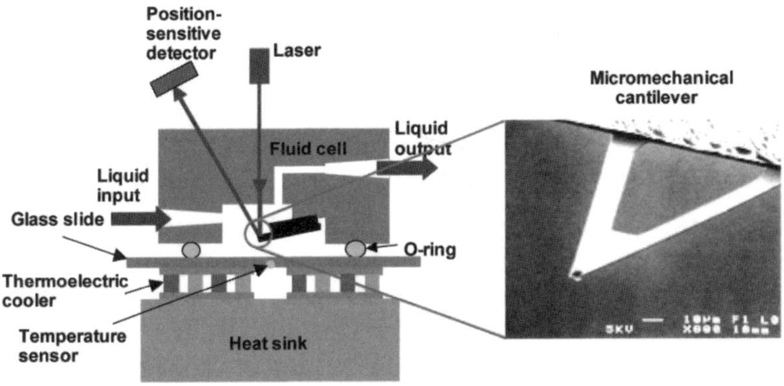

Figure 1.5. *Schematic diagram of an experimental setup using a MEMS cantilever for biosensing application. The scanning electron micrograph on the right shows the geometry of the gold-coated silicon nitride cantilever beam that is 200 µm long and 0.5 µm thick. Measurement of the cantilever deflection is carried out by reflecting a laser off the back of the cantilever and focusing it onto a position-sensitive detector [WU 01]*

The resonance of the structure can also be used as a sensing mechanism, bearing in mind that it requires active driving of the cantilever and feedback for frequency measurements. Piezoelectric thin films are often used for that purpose, although magnetic actuation [VAN 05] has also been applied. Cantilevered AT-cut quartz-crystal resonators have also been fabricated by deep reactive ion etching [LIN 05]. One major challenging issue is to improve the quality factor of the resonator. Q-factors of approximately 10^3 in the upper kilohertz frequency range in air enable a mass resolution in the picogram range [LUC 06]. Magnetic actuation and a closed-feedback loop can substantially enhance the quality factor [VID 03]. Another method for mass sensitivity to be improved is the use of higher modes [SAY 04] (as they exhibit higher resonant frequencies and Q-factor values).

However, in a liquid environment, which is the case of real-time biosensing applications, the out-of-plane vibration of a microstructure is strongly damped and results in an essentially reduced Q-factor (of a few units for cantilevered microstructures [BER 00]). One solution to tackle this issue consists of intentionally trapping air bubbles in the fluidic cell [AYE 07]. Another approach allows avoiding the out-of-plane vibration through the use of disc-shaped microstructures operated in rotational in-plane mode, thus exhibiting Q-factors as high as 5,800 in air and 100 in liquid [HO 05]. A real paradigm shift that has been proposed by the Manalis group consists of integrating the fluid circulation through channels buried in the sensors [BUR 07]. Although requiring multiple steps for fabricating the cantilever sensor, this case results in the resonator bearing the fluidic cell, opposite to approach commonly used (Figure 1.6). In this configuration, the cantilever vibrates in air or in vacuum (with no damping induced by a surrounding liquid) and the mass change occurs within the buried channels because of adsorption of the molecules of interest in liquid phase on the channel walls. Here, the signal is based on density differences between the attached molecule and the buffer.

1.4. One specific application of MEMS biosensors: detection of pathogen agents

After the terrorist incidents in 2001, an attack using a biological pathogen agent (BPA) became a conceivable event. Evidence of the destructiveness of such attacks and the associated fear they create have already been proven through episodes such as the sarin gas attacks by the Aum Shinrikyo cult in Tokyo in 1995, the use of salmonella in Oregon restaurants in 1984 by the Rajneeshee cult and the anthrax letter attacks soon after September 11. One of the problems with biological attacks is actually determining whether an attack has occurred. The difficulty arises because the initial symptoms after infection from BPAs are often difficult to distinguish from symptoms of infections due to more benign biological agents. The solution to this diagnosis issue is to use biosensing techniques, which can identify chemical markers from known biological agents.

a)

b)

c)

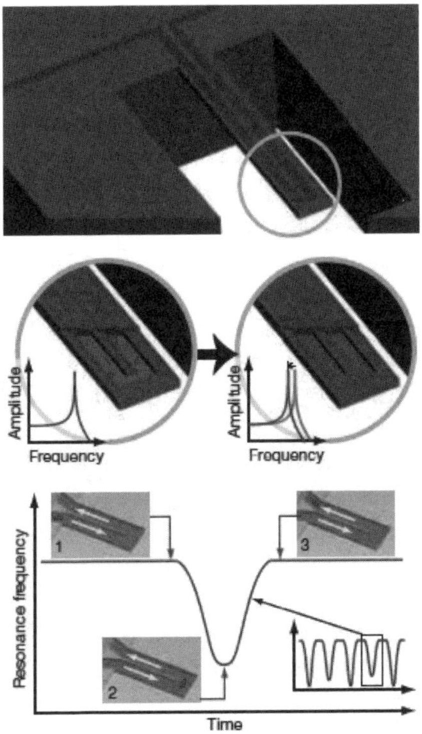

Figure 1.6. *Illustration of two mass measurement modes enabled by a fluid filled microcantilever [BUR 07]. a) A suspended microchannel translates mass changes into changes in resonance frequency. Fluid continuously flows through the channel and delivers biomolecules, cells or synthetic particles. Sub-femtogram mass resolution is attained by shrinking the wall and fluid layer thickness to the micrometer scale and by packaging the cantilever under high vacuum. b) While bound and unbound molecules both increase the mass of the channel, species that bind to the channel wall accumulate inside the device, and, as a result, their number can greatly exceed the number of free molecules in solution. This enables specific detection by way of immobilized receptors. c) In another measurement mode, particles flow through the cantilever without binding to the surface, and the observed signal depends on the position of particles along the channel (insets 1–3). The exact mass excess of a particle can be quantified by the peak frequency shift induced at the apex*

These agents are now well known and classified as category A agents by the Center for Disease Control and Detection (CDC) in the United States [BRO 01], including anthrax, tularemia, botulinum toxin, plague (aerosol version only), smallpox and hemorrhagic fever. They cover the main types of biological agents from gram-positive bacteria, which form spores (*Bacillus anthracis*, anthrax; *Yersinia pestis*, pneumonic plague), gram-negative bacteria (*Francisella tularensis*, tularemia) and toxins derived from bacterial species (botulinum toxin from *Clostridium botulinum*) and viruses (smallpox).

To develop biosensors for these analytes, it is of primary importance to consider the matrix in which the analyte will be found (for example air, water, food and person), the type of analyte (spore, bacteria, virus and toxin) and the possible features of the analyte (in order to detect it with an appropriate biorecognition element). Identification of such features requires some knowledge of the biological species and the forms in which they are found. Other considerations further concern the kind of measurement the biosensor has to perform, i.e. early identification of an infection ("detect-to-treat" systems) or providing a warning that a site is contaminated by BPAs to prevent infection ("detect-to-prevent" systems). The criteria for such biosensors are quite different. A detect-to-treat system must be able to identify a BPA from a biological sample within a few hours of infection. Hence detect-to-treat biosensing can be performed with analytical laboratory tools by specifically trained personnel. A detect-to-protect system must be able to provide a warning within a couple of minutes from, most frequently, an airborne sample without user intervention. Portability, reliability, fast response (real time), low detection limit and ruggedized packaging are thus critical requirements that correspond to the latter class of systems and that could be fulfilled by micromachined acoustic sensors.

Although there is currently no commercial "detect-to-protect" microsystem available, the literature of recent years abounds in papers where MEMS biosensors are fabricated and tested aiming at the detection of BPAs or simulants. A representative list of affinity biosensing concepts that are motivated toward studying detect-to-protect micromechanical devices is presented in Table 1.1 with some of their attributes.

Of course, none of the microsensors discussed above have left the laboratory, let alone satisfied the most important requirement, i.e. a detect-to-protect biosensor that could be worn by a soldier or a civilian in an environment where the possibility of exposure to BPAs is prevalent. There are still a number of impediments to the development of such a biosensor including (1) collecting and transferring an airborne sample to a confined environment where the biosensor can operate, (2) reducing the detection limit and obtaining faster responses and (3) achieving and thus demonstrating the actual integration of the entire biosensing platform. If the first criterion is quite specific to BPAs-related biosensors (but should be addressed for any application by deriving solutions offered in lab-on-chips or micro total analysis systems (µTAS)), and the second criterion is generic and must be satisfied for other biosensing applications ranging from pathology diagnosis to fundamental biophysics, the third criterion is an actual barrier to the commercial success of self-claimed miniaturized sensors. This means that future work should focus on the integration of actuation/sensing schemes, readout electronics and result display or delivery (for remote operation) to enable prototyping MEMS sensing platforms to possibly move beyond research laboratories.

Analytes	Detection limit	Assay time	Reagent free	Real samples	Comment
Bacillus subtilis spores	N/A	40 min	Yes	Yes	Array of eight cantilevers (four are used for detection and four for control purposes) operating in static mode [DHA 06]
Bacillus anthracis	333 spores/mL	30 min	Yes	Yes	Piezoelectric-excited millimeter-sized cantilever operating in resonant mode, in flow. The targeted spores are mixed with other bacterial strains [CAM 07a]
Bacillus anthracis	300 spores/mL	1 h	Yes	Yes	Piezoelectric-excited millimeter-sized cantilever operating in resonant mode, in flow [CAM 06a]
Bacillus anthracis	N/A	Few hours	Yes	Yes	Microcantilevers driven by thermally induced oscillations, detection of resonant frequency shift in air and water using a laser Doppler vibrometer [DAV 07]
Escherichia coli	100 cells/mL	Few minutes	Yes	Yes	Piezoelectric-excited millimeter-sized cantilever operating in resonant mode, in flow, sequential injection of the different materials [MAR 07]
Escherichia coli	50 cells/mL	2–6 h	Yes	Yes	Piezoelectric-excited millimeter-sized cantilever operating in resonant mode, in flow [CAM 07b]
Group A strepotococus	700 cells/mL	Few minutes	Yes	No	Piezoelectric-excited millimeter-sized cantilever operating in resonant mode, in flow [CAM 06b]
Salmonella enterica	10^6 cfu/mL (cfu = colony forming units)	Few minutes	Yes	Yes	Atomic force microscope setup, measurements performed in static mode, no circulating flow [WEE 03]
Vaccinia virus	2 mg/mL	Several hours	Yes	Yes	Piezoresistive cantilevers operating in static mode, measurements performed in aerosol and liquid samples [GUN 03]

Table 1.1. *A representative list of affinity biosensing concepts implemented on detect-to-protect micromechanical devices ("reagent free" refers to cases where no extra reagent is needed for the measurement, regardless of the sample preparation consideration)*

1.5. Bibliography

[AYE 07] AYELA C., NICU L., "Micromachined piezoelectric membranes with high nominal quality factors in Newtonian liquid media: a Lamb's model validation at the microscale", *Sensors and Actuators Chemical*, vol. B123, pp. 860–868, 2007.

[AYE 08] AYELA C., ALAVA T., LAGRANGE D., *et al.*, "Electronic scheme for multiplexed dynamic behavior excitation and detection of piezoelectric silicon-based micromembranes", *IEEE Sensors Journal*, vol. 8, pp. 210–217, 2008.

[BAS 04] BASHIR R., "BioMEMS: state-of-the-art in detection, opportunities and prospects", *Advanced Drug Delivery Reviews*, vol. 56, pp. 1565–1586, 2004.

[BAS 06] BASHIR R., WERELEY S., *Biomolecular Sensing, Processing, and Analysis in BioMEMS and Biomedical Nanotechnology*, Springer, Berlin, 2006.

[BER 00] BERGAUD C., NICU L., "Viscosity measurements based on experimental investigations of composite cantilever beam eigenfrequencies in viscous media", *Review of Scientific Instruments*, vol. 71, pp. 2487–2491, 2000.

[BES 03] BESTEMAN K., LEE J., WIERTZ F.G.M., *et al.*, "Enzyme-coated carbon nanotubes as single-molecule biosensors", *Nano Letters*, vol. 3, pp. 727–730, 2003.

[BHA 99] BHATIA S.N., CHEN C.S., "Three-dimensional photopatterning of hydrogels containing living cells", *Biomedical Microdevices*, vol. 2, pp. 131–144, 1999.

[BIE 96] BIER F.F., EHRENTREICH-FORSTER E., BAUER C.G., *et al.*, "High sensitive competitive immunodetection of 2,4-dichlorophenoxyacetic acid using enzymatic amplification with electrochemical detection", *Fresenius Journal of Analytical Chemistry*, vol. 354, pp. 861–865, 1996.

[BOR 97] BORKHOLDER D.A., BAO J., MALUF N.I., *et al.*, "Microelectrode arrays for stimulation of neural slice preparations", *Journal of Neuroscience Methods*, vol. 77, pp. 61–66, 1997.

[BRA 06] BRAUN T., BACKMANN N., VOGTLI M., *et al.*, "Conformational change of bacteriorhodopsin quantitatively monitored by microcantilever sensors", *Biophysical Journal*, vol. 90, pp. 2970–2977, 2006.

[BRO 01] BROUSSARD L.A., "Biological agents: weapons of warfare and bioterrorism", *Molecular Diagnosis*, vol. 6, pp. 323–333, 2001.

[BUE 93] BUERK D.G., *Biosensors: Theory and Applications*, Technomic Publishing Company, 1993.

[BUR 07] BURG T.P., GODIN M., KNUDSEN S.M, *et al.*, "Weighing of biomolecules, single cells and single nanoparticles in fluid", *Nature*, vol. 446, pp. 1066–1069, 2007.

[CAM 06a] CAMPBELL G. A., MUTHARASAN R., "Detection of Bacillus anthracis spores and a model protein using PEMC sensors in a flow cell at 1 mL/min", *Biosensors Bioelectronics*, vol. 22, pp. 78–85, 2006.

[CAM 06b] CAMPBELL G.A., MUTHARASAN R., "PEMC sensor's mass change sensitivity is 20 pg/Hz under liquid immersion", *Biosensors Bioelectronics*, vol. 22, pp. 35–41, 2006.

[CAM 07a] CAMPBELL G.A., UKNALIS J., TU S.I., *et al.*, "Detect of Escherichia coli O157: H7 in ground beef samples using piezoelectric excited millimeter-sized cantilever (PEMC) sensors", *Biosensors Bioelectronics*, vol. 22, pp. 1296–1302, 2007.

[CAM 07b] CAMPBELL G.A., MUTHARASAN R., "Method of measuring Bacillus anthracis spores in the presence of copious amounts of Bacillus thuringiensis and Bacillus cereus", *Analytical Chemistry,* vol. 79, pp. 1145–1152, 2007.

[CHI 07] CHINOWSKY T.M., SOELBERG S.D., BAKER P., *et al.*, "Portable 24-analyte surface plasmon resonance instruments for rapid, versatile biodetection", *Biosensors Bioelectronics*, vol. 22, pp. 2268–2275, 2007.

[CUI 01] CUI Y., WEI Q., PARK H., *et al.*, "Nanowire nanosensors for highly sensitive and selective detection of biological and chemical species", *Science,* vol. 293, pp. 1289–1292, 2001.

[DAV 07] DAVILA A.P., JANG J., GUPTA A.K., *et al.*, "Microresonator mass sensors for detection of Bacillus anthracis Sterne spores in air and water", *Biosensors Bioelectronics*, vol. 22, pp. 3028–3035, 2007.

[DHA 06] DHAYAL B., HENNE W. A., DOORNEWEERD D.D., *et al.*, "Detection of Bacillus subtilis spores using peptide-functionalized cantilever arrays", *Journal of the American Chemical Society*, vol. 128, pp. 3716–3721, 2006.

[DRU 03] DRUMMOND T.G., HILL M.G., BARTON J.K., "Electrochemical DNA sensors", *Nature Biotechnology*, vol. 21, pp. 1192–1199, 2003.

[FRI 00] FRITZ J., BALLER M.K., LANG H.P., *et al.*, "Translating biomolecular recognition into nanomechanics", *Science*, vol. 288, pp. 316–318, 2000.

[GIZ 02] GIZELI E., LOWE C.R., *Biomolecular Sensors*, Taylor & Francis, 2002.

[GRE 91] GREEN M.J., BARRANCE D.J., HILDITCH P.I., "Electrochemical immunoassay", in PRICE C.P., NEWMAN D.J. (eds), *Principles and Practice of Immunoassay*, Macmillan, London, 1991.

[GRO 77] GROSS G.W., RIESKE E., KREUTZBERG G.W., *et al.*, "New fixed-array multi-electrode system designed for long-term monitoring of extracellular single unit neuronal-activity in vitro", *Neuroscience Letters,* vol. 6, pp. 101–105, 1977.

[GUI 81] GUILBAULT G.G., "Analysis of environmental pollutants using a piezoelectric crystal detector", *International Journal of Environmental Analytical Chemistry*, vol. 10, pp. 89–98, 1981.

[GUN 03] GUNTER R.L., DELINGER W.G., MANYGOATS K., *et al.*, "Viral detection using an embedded piezoresistive microcantilever sensor", *Sensors and Actuators Physical*, vol. A107, pp. 219–224, 2003.

[HAL 90] HALL E.A.H., *Biosensors*, Open University Press, 1990.

[HIN 95] HINTSCHE R., KRUSE C., UHLIG A., PAESCHKE M., LISEC T., SCHNAKENBERG U., WAGNER B., "Chemical microsensor systems for medical applications in catheters", *Sensors and Actuators Chemical*, vol. B27, pp. 471–473, 1995.

[HO 95] HO W.O., ATHEY D., MCNEIL C.J., "Amperometric detection of alkaline-phosphatase activity at a horseradish-peroxidase enzyme electrode based on activated carbon – potential application to electrochemical immunoassay", *Biosensors Bioelectronics*, vol. 10, pp. 683–691, 1995.

[HÖÖ 97] HÖÖK F., Development of a novel QCM technique for protein adsorption studies, PhD Thesis, Chalmers University of Technology, Gothenburg, Sweden, 1997.

[HÖÖ 98a] HÖÖK F., RODAHL M., BRZEZINSKI P., *et al.*, "Energy dissipation kinetics for protein and antibody-antigen adsorption under shear oscillation on a quartz crystal microbalance", *Langmuir*, vol. 14, pp. 729–734, 1998.

[HÖÖ 98b] HÖÖK F., RODAHL M., KASEMO B., *et al.*, "Structural changes in hemoglobin during adsorption to solid surfaces: effects of pH, ionic strength, and ligand binding", *Proceedings of the National Academy of Sciences of the United States of America*, vol. 95, pp. 12271–12276, 1998.

[JAN 00] JANSHOFF A., GALLA H.J., STEINEM C., "Piezoelectric mass-sensing devices as biosensors – an alternative to optical biosensors?", *Angewandte Chemie International Edition*, vol. 39, pp. 4004–4032, 2000.

[JOH 91] JOHNSSON B., LÖFÅS S., LINDQUIST G., "Immobilization of proteins to a carboxymethyldextran-modified gold surface for biosepcific interaction analysis in surface-plasmon resonance sensors", *Analytical Biochemistry*, vol. 198, pp. 268–277, 1991.

[JON 91] JONSSON U., FAGERSTAM L., IVARSSON B., *et al.*, "Real-time biospecific interaction analysis using surface-plasmon resonance and a sensor chip technology", *Biotechniques*, vol. 11, pp. 620–627, 1991.

[KAN 85] KANAZAWA K.K., "Parasitic drag on a quartz resonator in liquid", *Journal of Electrochemical Society*, vol. 132, p. 368, 1985.

[KAS 78] KASEMO B., TÖRNQVIST E., "Quartz crystal microbalance measurements of O-Ti and CO-Ti atom ratios on very thin Ti films", *Surface Science*, vol. 77, pp. 209–218, 1978.

[KAS 80] KASEMO B., TÖRNQVIST E., "Weighing fractions of monolayers – application to the adsorption and catalytic reactions of H_2, CO and O_2 on Pt", *Physical Review Letters*, vol. 44, pp. 1555–1558, 1980.

[KIN 64] KING JR W.H., "Piezoelectric sorption detector", *Analytical Chemistry*, vol. 36, pp. 1735–1739, 1964.

[KOV 98] KOVACS G.T.A., *Micromachined Transducers Sourcebook*, WCB/McGraw-Hill, Boston, MA, 1998.

[KRE 68] KRETSCHMANN E., RAETHER H., "Radiative decay of non-radiative surface plasmons excited by light", *Zeitschrift für Naturforschung Part A – Astrophysik Physic und Physikalische Chemie*, vol. A23, pp. 2135–2136, 1968.

[LEH 01] LEHMANN M., BAUMANN W., B M., *et al.*, "Simultaneous measurement of cellular respiration and acidification with a single CMOS ISFET", *Biosensors Bioelectronics,* vol. 16, pp. 195–201, 2001.

[LIE 83] LIEDBERG B., NYLANDER C., LUNDSTRÖM I., "Surface-plasmon resonance for gas detection and biosensing", *Sensors and Actuators*, vol. 4, pp. 299–304, 1983.

[LIN 05] LIN Y.C., ONO T., ESASHI M., "Resonating quartz-crystal cantilever for force sensing", *Proceedings of the 13th International Conference on Solid-State Sensors Actuators and Microsystems* (*Transducers '05*), Seoul, South Korea, 593–596, 5–9 June 2005.

[LÖF 90] LÖFAS S., JOHNSSON B., "A novel hydrogel matrix on gold surfaces in surface-plasmon resonance sensors for fast and efficient covalent immobilization of ligands", *Journal of the Chemical Society – Chemical Communications*, vol. 21, pp. 1526–1528, 1990.

[LUC 03] LUCKLUM R., HAUPTMANN P., "Acoustic microsensors – the challenge behind microgravimetry", *Analytical and Bioanalytical Chemistry*, vol. 384, pp. 667–682, 2006.

[MAR 07] MARALDO D., RIJAL K., CAMPBELL G., *et al.*, "Method for label-free detection of femtogram quantities of biologics in flowing liquid samples", *Analytical Chemistry*, vol. 79, pp. 2762–2770, 2007.

[MEA 96] MEADOWS D., "Recent developments with biosensing technology and applications in the pharmaceutical industry", *Advanced Drug Delivery Reviews*, vol. 21, pp. 179–189, 1996.

[MUH 03] MUHAMMAD T.Z., ALOCILJA E.C., "A conductometric biosensor for biosecurity", *Biosensors Bioelectronics*, vol. 18, pp. 813–819, 2003.

[MUK 05] MUKHOPADHYAY R., SUMBAYEV V.V., LORENTZEN M., *et al.*, "Cantilever sensor for nanomechanical detection of specific protein conformations", *Nano Letters*, vol. 5, pp. 2385–2388, 2005.

[NAI 05] NAIMUSHIN A.N., SPINELLI C.B., SOELBERG S.D., *et al.*, "Airborne analyte detection with an aircraft-adapted surface plasmon resonance sensor system", *Sensors and Actuators Chemical*, vol. B 104, pp. 237–248, 2005.

[NIW 95] NIWA O., "Electroanalysis with interdigitated array microelectrodes", *Electroanalysis*, vol. 7, pp. 606–613, 1995.

[NYL 82] NYLANDER C., LIEDBERG B., LIND T., "Gas detection by means of surface-plasmon resonance", *Sensors and Actuators*, vol. 3, pp. 79–88, 1982.

[OTT 68] OTTO A., "Excitation of nonradiative surface plasma waves in silver by method of frustrated total reflection", *Zeitschrift für Physik*, vol. 216, pp. 398–410, 1968.

[PEP 86] PEPPAS N.A., *Hydrogels in Medicine and Pharmacy*, CRC, Boca Raton, FL, 1986.

[POC 78] POCKRAND I., SWALEN J.D., GORDON J.G., *et al.*, "Surface plasmon spectroscopy of organic monolayer assemblies", *Surface Science*, vol. 74, pp. 237–244, 1978.

[POL 00] POLLA D.L., ERDMAN A.G., ROBBINS W.P., *et al.*, "Microdevices in medicine", *Annual Review in Biomedical Engineering*, vol. 2, pp. 551–576, 2000.

[RAB 03] RABE J., BUTTGENBACH S., SCHRODER J., *et al.*, "Monolithic miniaturized quartz microbalance array and its application to chemical sensor systems for liquids", *IEEE Sensors Journal*, vol. 3, pp. 361–368, 2003.

[RAS 03] RASMUSSEN P.A., THAYSEN J., HANSEN O., *et al.*, "Optimised cantilever biosensor with piezoresistive read-out", *Ultramicroscopy*, vol. 97, pp. 371–376, 2003.

[RIC 04] RICHTER R., The formation of solid-supported lipid membranes and two-dimensional assembly of proteins. A study combining atomic force microscopy and quartz crystal microbalance with dissipation monitoring, PhD Thesis, University of Bordeaux, France, 2004.

[RUD 96] RUDEL U., GESCHKE O., CAMMANN K., "Entrapment of enzymes in electropolymers for biosensors and graphite felt based flow-through enzyme reactors", *Electroanalysis*, vol. 8, pp. 1135–1139, 1996.

[SAR 99] SARKAR P., TURNER A.P.F., "Application of dual-step potential on single screen-printed modified carbon paste electrodes for detection of amino acids and proteins", *Fresenius Journal of Analytical Chemistry,* vol. 364, pp. 154–159, 1999.

[SAU 59] SAUERBREY G., "Verwendung von scwhinquarzen zur wagung dunner schichten und zur mikrowagung", *Zeitschrift fur Physik*, vol. 155, pp. 206–222, 1959.

[SAY 04] SAYA D., NICU L., GUIRARDEL M., *et al.*, "Mechanical effect of gold nanoparticles labeling used for biochemical sensor applications: a multimode analysis by means of SiNx micromechanical cantilever and bridge mass detectors", *Review of Scientific Instruments*, vol. 75, pp. 3010–3015, 2004.

[SCH 96] SCHNAKENBERG U., LISEC T., HINTSCHE R., *et al.*, "Novel potentiometric silicon sensor for medical devices", *Sensors and Actuators Chemical*, vol. B34, pp. 476–480, 1996.

[SEO 05] SEO J.H., BRAND O., "Novel high Q-factor resonant microsensor platform for chemical and biological applications", *13th International Conference on Solid-State Sensors, Actuators and Microsystems (Transducers '05*), Seoul, South Korea, pp. 247–250, 5–9 June 2005.

[SEP 06] SEPÚLVEDA B., SÁNCHEZ DEL RÍO J., MORENO M., *et al.*, "Optical biosensor microsystems based on the integration of highly sensitive Mach-Zehnder interferometer devices ", *Journal of Optics A: Pure and Applied Optics,* vol. 8, pp. 561–566, 2006.

[SHU 94] SHUL'GA A.A., SOLDATKIN A.P., EL'SKAYA A.V., *et al.*, "Thin-film conductometric biosensors for glucose and urea determination", *Biosensors Bioelectronics*, vol. 9, pp. 217–223, 1994.

[STE 97] STEINSCHADEN A., ADAMOVIC D., JOBST G., *et al.*, "Miniaturised thin film conductometric biosensors with high dynamic range and high sensitivity", *Sensors and Actuators Chemical*, vol. B44, pp. 365–369, 1997.

[SUZ 01] SUZUKI H., ARAKAWA H., KARUBE I., "Fabrication of a sensing module using micromachined biosensors", *Biosensors Bioelectronics,* vol. 16, pp. 725–733, 2001.

[TAK 97] TAKADA K., DÍAZ D.J., ABRUÑA H.D., *et al.*, "Redox-active ferrocenyl dendrimers: thermodynamics and kinetics of adsorption, in-situ electrochemical quartz crystal microbalance study of the redox process and tapping mode AFM imaging", *Journal of American Chemical Society*, vol. 119, pp. 10763–10773, 1997.

[VAN 05] VANCURA C., RUEGG M., LI Y., *et al.*, "Magnetically actuated complementary metal oxide semiconductor resonant cantilever gas sensor systems", *Analytical Chemistry*, vol. 77, pp. 2690–2699, 2005.

[VAN 07] VANCURA C., LI Y., LICHTENBERG J., *et al.*, "Liquid-phase chemical and biochemical detection using fully integrated magnetically actuated complementary metal oxide semiconductor resonant cantilever sensor systems", *Analytical Chemistry*, vol. 79, pp. 1646–1654, 2007.

[VID 03] VIDIC A., THEN D., ZIEGLER C., "A new cantilever system for gas and liquid sensing", *Ultramicroscopy*, vol. 97, pp. 407–416, 2003.

[WEE 03] WEEKS B.L, CAMARERO J., NOY A., *et al.*, "A microcantilever-based pathogen detector", *Scanning*, vol. 25, pp. 297–299, 2003.

[WU 01] WU G., DATAR R., HANSEN K., THUNDAT T., *et al.*, "Bioassay of prostate-specific antigen (PSA) using microcantilevers", *Nature Biotechnology*, vol. 19, pp. 856–860, 2001.

[XIA 98] XIA Y., WHITESIDES G.M., "Soft lithography", *Annual Review of Material Sciences*, vol. 28, pp. 153–184, 1998.

[YAZ 98] YAZDI N., AYAZI F., NAJAFI K., "Micromachined inertial sensors", *Proceedings of the IEEE*, vol. 86, pp. 1640–1659, 1998.

[ZIM 01] ZIMMERMANN B., LUCKLUM R., HAUPTMANN P., *et al.*, "Electrical characterisation of high-frequency thickness-shear-mode resonators by impedance analysis", *Sensors and Actuators Chemical*, vol. B76, pp. 47–57, 2001.

2

Bioreceptors and Grafting Methods

Surface functionalization is a critical step in the fabrication process of a biosensor. Indeed, physical sensors such as microelectromechanical system (MEMS) sensors can be rendered specifically sensitive to the presence of biological species provided they are adequately biofunctionalized, i.e. the surfaces of the MEMS structure are coated with a biosensitive layer. The biosensitive layer consists of bioreceptors that are typically either biocatalytic, affinitive, nucleic acid related or biomimetic.

To create a molecular recognition interface that involves localization and attachment of the bioreceptors near to or onto the transducer surface, several immobilization techniques are available, e.g. physical adsorption, entrapment or covalent coupling. The choice of the optimum technique for a specific application depends on various factors such as the transduction principle, the surface chemistry of the sensor, the nature of the biological receptor, the nature of the analyte to be detected and the way the biosensor is to be used. Along with the use of a particular grafting strategy, it is mandatory to rely on techniques that allow achieving localized surface immobilization appropriate to the functionalization of miniaturized biosensors.

The aim of this chapter is to address the issues raised by the biofunctionalization of micromachined sensors and to present the different solutions that can be envisioned according to the targeted application. It is organized as follows: first, we present the different classes of bioreceptors used to capture/detect the target biological entities and the typical molecules used as antifouling agents; second, we list the chemical/physical means used to anchor the biomolecular receptors to the sensor surface. The various techniques available to pattern the biological layer are discussed in Chapter 3 of this book.

2.1. Types of bioreceptor

The biological recognition element, probe or bioreceptor is the most crucial component of the biosensor device. The bioreceptor is responsible for the selective recognition of the analyte to be detected and is thus the key to specificity. Biological receptors can generally be classified into three categories: biocatalytic (as typified by enzymes), immunological or affinitive (of which the antibodies are the best-known examples) and nucleic acid related (e.g. deoxyribonucleic acid (DNA)). In addition to these types of receptors, we may use biomimetic receptors [VOD 00], i.e. engineered receptors, among which molecular imprinted polymers (MIPs), even if non-biological, are highlighted here since they offer interesting alternative ways to fabricate MEMS biosensors. Finally, it is important to note that some biosensors have been developed using cellular systems [KIM 06], or cellular components [VOD 00]. However, these bioreceptors are not discussed here since their current use with MEMS sensors is limited, probably due to integration and viability issues.

2.1.1. *Catalytic receptors*

Enzymes are mostly protein molecules, i.e. long chains of amino acids that are structured so as to confer a remarkable ability for catalyzing specific reactions. However, this structure also limits their functional stability. Compared to non-biological catalysts, enzymes are 10^8–10^{13} times more active and are capable of producing hundreds of thousands of molecules per second [MAD 93].

Enzymes have the longest tradition in the field of biosensors. Since 1997, more than 2,000 articles have published in the literature on enzyme-based biosensors. These biosensors primarily rely on two operational mechanisms. They can be used as bioreceptors based on their specific binding capabilities or according to their catalytic transformation of a species into a detectable form [VOD 00]. Most reports in the literature show that they have been employed according to their catalytic activity [TOT 01].

Horseradish peroxidase, alkaline phosphatase and glucose oxidase are three enzymes that have been employed in most biosensor studies [ROG 98, LAS 00].

The detection limit of these biosensors is mainly determined by the activity of the enzyme, which can be described by the Michaelis–Menten equation. The major limitation of enzyme-based biosensors is the stability of the enzyme, which depends on various conditions such as temperature and pH [ROG 98, LAS 00]. The ability to maintain enzyme activity for a long period of time still remains a major obstacle [LIU 12]. Another issue that governs the success of an enzyme-based sensor depends primarily on the contact between the enzyme and electrode surface [WAN 00]. Despite these pitfalls, the enzyme-based sensor is still the most commonly used biosensor, and this is largely due to the need for monitoring glucose in blood [TOT 01]. Still, some recent studies have shown that enzyme-based biosensors can be used to detect very low levels (i.e. $\sim10^{-17}$ M) of pesticides [PRI 04].

2.1.2. *Affinity receptors*

The affinity bioreceptors bind to ligands with high specificity, i.e. with binding constants of 10^{9}–10^{12} M^{-1}, and do not exhibit catalytic activity. The first consequence for sensing application is that affinity receptors are more suited to "one-shot" detection rather than monitoring applications, since the binding is essentially irreversible. Changing the pH of the environment to acidic values (typically below 2) can break the binding complex, although this tends to reduce the affinity and specificity of the receptor. The high binding constant also favors the detection of a very small amount of analyte with high specificity. Affinity receptors are suited to analyte concentrations of 10^{-6}–10^{-9} M [STE 90], as compared with 10^{-3}–10^{-6} M for catalytic systems. The detection of concentrations as low as 10^{-15} M has even been demonstrated recently [SHA 13].

Of the affinity receptors concerned with biosensing applications, the immunoreceptors (antibody/antigen) are the dominant type. Antibodies are highly selective, chemical attractor molecules (glycoproteins) that are produced by mammalian immunological systems in response to the introduction of an antigen. An antigen is simply any foreign substance (molecule, hormone, virus, bacteria, etc.) that elicits an immune response, thus antibody formation, and reacts specifically with that antibody. Immunogenicity, which is the ability of the antigen to provoke an immune response, is not an intrinsic property of the antigen but a relational property that depends on the gene repertoire and regulatory mechanisms of the immunized host and that has no meaning outside the context of the host [GIZ 02]. For instance, human serum albumin (HSA) is an antigenic protein that is immunogenic in mice but not normally in humans because of the regulatory mechanism known as immunological tolerance (the opposite being the starting point of autoimmune diseases).

Since the antibody–antigen reaction is highly specific, each can be used as a specific chemical detector for the other. Antibodies that are typically represented schematically as Y-shaped structures (Figure 2.1) consist of two identical Fab (fragment, antigen binding) portions hinged to an Fc (fragment, crystallizable) part. The variable regions of the Fab fragments are where the amino acids are organized to produce a binding site for the specific antigen, two binding sites per antibody. The Fc fragment of the antibody does not combine with the antigen but contains carboxyl-terminated amino acids that allow linkage to solid substrates (e.g. that of transducers).

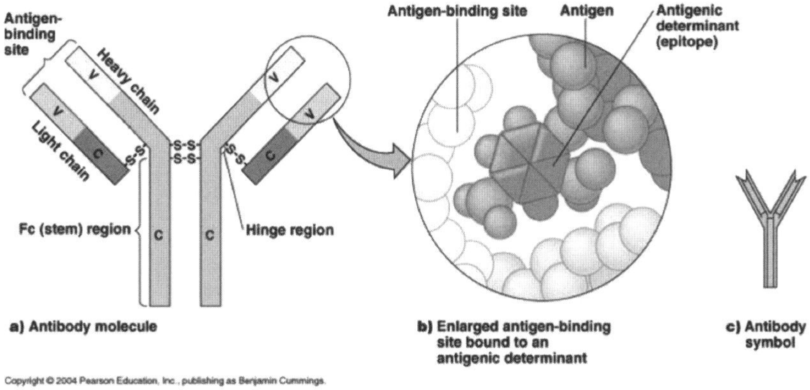

Figure 2.1. *Structure of an antibody*

Immunoassays have become a standard technique in clinical chemistry, since such biochemical tests can be performed by sophisticated automated instrument and large numbers of samples can be analyzed in short time frames [HEN 98]. Immunosensors have been derived from the standard immunoassay approach [HOC 97], and either the antigen or the antibody can be attached onto the sensor surface, although the latter approach has been used most often [STE 00]. Optical [KUS 98, GOM 00, KOC 00] and electrochemical [LEE 00, SUS 03] detection methods are most frequently used in immunosensors, though since the recent commercialization of the quartz crystal microbalance-dissipative (QCM-D) system by the Q-Sense company (Gothenburg, Sweden), the piezoelectric detection has become a promising immunoassay technique [AYE 07a]. A common problem with immunosensors is that they are not completely reversible so that only a single immunoassay can be performed. However, some recent research efforts have been directed toward the development of renewable antibody surfaces [CAN 04].

2.1.3. *Nucleic acid-based receptors*

Nucleic acids are linear biopolymers that can encode, transmit and express genetic information. Composed of multiple nucleotides and found in single or double strands, the nucleic acid family includes DNA, ribonucleic acid (RNA) and their related engineered molecules, such as oligonucleotides (short single-stranded DNA or RNA molecules), peptide nucleic acids (PNA) [NIE 91], locked nucleic acid (LNA) [KOS 98] and aptamers [SAV 10].

In DNA sensors, the recognition is based on the formation of stable hydrogen bonds between two nucleic acid strands. The bonding between nucleic acids takes place at regular (nucleotide) intervals along the length of the duplex. The specificity of nucleic acid probes relies on the ability of different nucleotides to form bonds only with an appropriate counterpart. An important property of DNA is that the nucleic acid ligands can be denatured to reverse the binding and thus to regenerate the probes by controlling pH and buffer ion concentration or by heating the sample [IVN 99]. For DNA/RNA detection, PNA can also be used as a biorecognition element. Since the backbone of PNA does not contain any charged phosphate groups, it is capable of binding very strongly to complementary oligonucleotide sequences [VOD 00].

The detection of specific DNA sequences provides the fundamental basis for detecting a wide variety of microbial and viral pathogens and there has already been a significant amount of work on the development and application of DNA sensors for the testing of virus infections [WAN 97]. In essence, the technology relies on the immobilization of a short (20–40 mer) synthetic oligomer (single-stranded DNA), whose sequence is complementary to the target of interest. Exposure of the sensor to a sample containing the target results in the formation of the hybrid on the surface, and various transduction methods (i.e. optical, electrochemical and piezoelectric) have been used to detect duplex formation [YAN 97a]. However, it is important to note that so far, relatively few DNA biosensor studies have been carried out in real complex biological samples [GOO 02].

More than 700 papers have appeared in the literature, since 1997, on the development of nucleic acid biosensors with targeted applications possibly ranging from expression analysis to disease diagnostic and mutation detection [CAM 04]. Almost all reported works on DNA biosensors have used relatively short synthetic oligonucleotides for detecting target DNAs of about the same length. Most reports have immobilized DNA in the form of a self-assembled monolayer (SAM) onto a gold surface using thiol chemistry. However, in some cases, binding of the oligonucleotide probe to the sensing surface is achieved by using the biotin/avidin interaction immobilization techniques.

2.1.4. *Molecularly imprinted polymers*

Even if at first the concept of a biosensor was likely to revolutionize point-of-care analysis, the pace of biosensor development has been slow and only a very small proportion of the perceived market has been filled. Many of the key issues are directly related to the stability of the biological macromolecules used at the heart of the biosensor. Thus, problems such as unpredictable shelf-life and stability of biological receptors, poor inter-batch reproducibility and availability, difficulties in incorporating biomolecules into sensor platforms, environmental intolerance (e.g. pH, temperature, ionic strength, organic solvents) and poor engineering characteristics have to be effectively addressed.

Molecular imprinting is a technology that could potentially address some of the key issues listed above as an alternative solution to biological receptors. Molecular imprinting consists of the design and synthesis of biomimetic receptors (artificial macromolecules) capable of binding a target molecule with similar affinities and specificities to their natural counterparts [YE 04]. In this technique, a target molecule (acting as a molecular template) is used to direct the assembly of specific binders trapped in a polymer matrix formed by a polymerization step (Figure 2.2). Besides the simplicity in separating MIPs from the soluble template (that allows easy recovering and using them as artificial immobilized antibodies, small molecules or enzyme mimics), MIPs display a higher chemical and physical stability compared to biomacromolecules that make them very attractive for applications covering biochemical analysis, separation and catalysis.

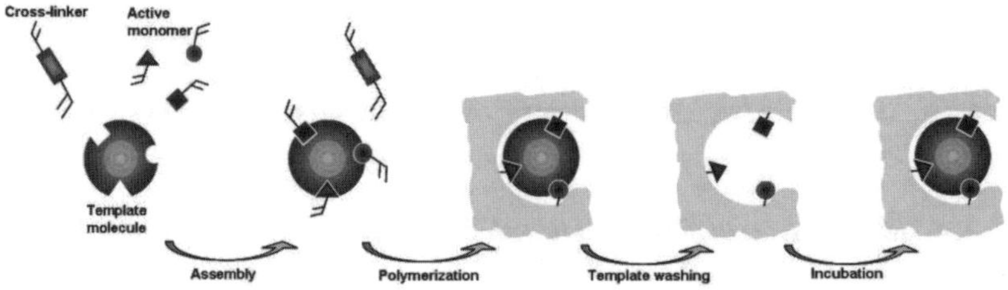

Figure 2.2. *Schematics of the molecular imprinting principle (courtesy of C. Ayela and K. Haupt)*

Since 1993, the ISI Web of Knowledge Database has registered more than 1,100 publications dealing with MIPs (http://portal.isiknowledge.com/), taking roots in a report by the Mosbach group on the development of a MIP-based immunoassay against theophylline and diazepam [VLA 93]. Following this, MIPs have been used as substitutes for antibodies in radioimmunoassays for drugs, showing strong binding to the target analytes (with dissociation constants as low as 10^{-6}–10^{-9} M) and cross-reactivity profiles similar to those of antibodies [AND 95, RAM 97]. MIPs were later used to develop assay systems for other compounds as well, such as herbicides [MUL 95, HAU 98].

Another aspect in assay development is the potential use of MIPs in automated systems for unattended monitoring. In a recent paper [SUR 01], Surugiu *et al.* have presented a flow-injection ELISA-type MIP assay using polymer microspheres coated glass capillary. A photomultiplier tube was used for the detection of the herbicide molecule 2,4-dichlorophenoxyacetic acid with concentrations ranging from 0.5 ng/mL to 50 µg/mL (2.25 nM to 225 µM), making the system one of the most sensitive MIP-based assays reported so far. Though non-exhaustive, the above listed successful demonstrations of the use of MIPs for assays, sensors and very recently for drug monitoring [HIL 05] reveal the great potential of this technology and its combination with acoustic microsensors [AYE 07b].

2.2. Immobilization strategies

One key factor in the design of biosensors is the development of immobilization technologies for stabilizing bioreceptor molecules and tethering them to surfaces (Figure 2.3) [COL 97]. The choice of the most appropriate grafting strategy among the ones listed in this section is governed by several elementary issues/questions to be addressed before adopting the final design of the biosensor:

– Will the sensor be regenerated and reused or dedicated to one-shot measurements?

– What is the nature of the biological receptor (hydrophobic/hydrophilic, large/small, biological/polymeric, etc.)?

– Are the bioreceptors sensitive to handling and, as a main concern, what are the storing constraints (given the fact that the bioreceptors must be kept functional during immobilization and until the recognition takes place at the surface of the biosensor)?

– Is the sensor intended as a proof of principle or a research tool, or is it planned for commercialization?

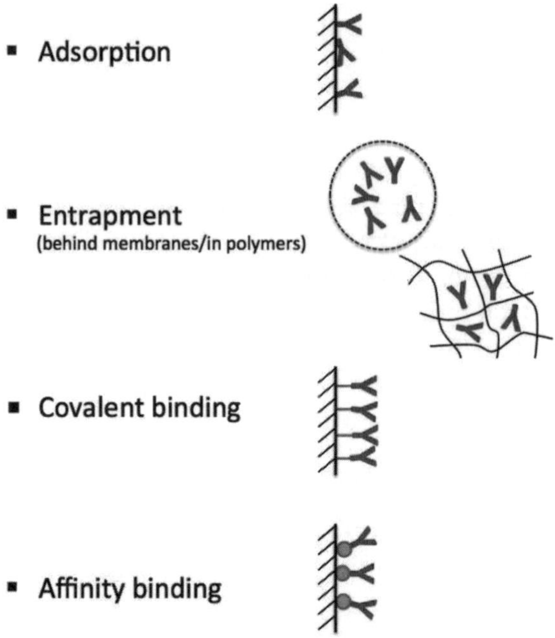

Figure 2.3. *Examples of some approaches used to immobilize biomolecules onto a surface*

As stated by Andrew G. Mayes [GIZ 02], "arriving at the correct compromise for any particular situation is a matter of knowledge, experience and careful consideration of the known parameters in relation to the types of issues outlined above".

2.2.1. *Adsorption and antifouling strategies*

Physical adsorption is the most straightforward way to immobilize biomolecules onto a solid surface, especially in the case of proteins. In general, proteins adsorb natively to hydrophobic surfaces; this phenomenon is often followed by a thermodynamically driven unfolding of the protein structure, in order to increase the amount of a hydrophobic polypeptide chain (usually from the protein interior) in contact with the surface [WAH 91]. Though this leads to denaturation (and deactivation) of most proteins, antibodies are rather resistant since they have a very rigid tertiary structure.

The main concern regarding hydrophobic surfaces deals, however, with the non-specific binding issue. Indeed, strong adsorption means that any kind of protein present in solution tends to bind to the surface of and around the biosensor. Not only hydrophobic but also charged surfaces may non-specifically bind proteins due to ion-pair interactions between immobilized charges and ionized surface groups on the protein.

In body fluids, such as blood, for instance, since the concentration of plasma proteins (e.g. albumin) is at least six orders of magnitude higher than the minimum level of detection of biomarkers (e.g. PSA) that is required for a disease diagnosis [AND 02], non-specific adsorption can have devastating effects onto the measurement and thus the analysis result. Indeed, nonetheless non-specific adsorption of an analyte can lead to high background signals that cannot be differentiated from the intended specific binding, but non-specific adsorption of target molecules around the sensors (and not on its active surface) can also impede or delay the results if only a few copies of the targets are present in the sample (i.e. at ultra-low concentration). Thus, the non-specific adsorption issue needs to be taken into account and addressed when designing a biosensor.

To prevent non-specific binding and denaturation of proteins from occurring, surfaces most closely mimicking an aqueous solution environment (i.e. uncharged – so it does not ionize under physiological conditions, and possibly rich in hydroxyl groups – which favors hydrophilicity) have to be engineered. Thus, surfaces best suited for (covalent) immobilization should probably contain a mixture of OH and poly/oligo (ethylene glycol) chains [YAN 97b] (known for their repellent effect regarding proteins) in the form of SAMs and sufficient specific functional groups used to achieve the required degree of surface derivatization.

In the following section, the various approaches available as a first step to chemically modify a surface with the aim of the final attachment of a desired molecule will be discussed for the most common type of materials used to fabricate miniaturized transducers.

2.2.1.1. *Gold*

Gold is the most popular noble metal used in the fabrication of biosensors (either as an electrode material in electrochemical and acoustic sensors or as an integral part of most surface plasmon resonance (SPR) sensor devices). The main advantage of using gold arises from the following properties:

– Inert nature (it does not oxidize in air).

– Highly compatible with most semiconductor manufacturing processes.

– Allows the formation of SAMs at its surface [NUZ 83].

Thiol (R–SH), sulfide (R–S–R) and disulfide (R–S–S–R) molecules all self-assemble on gold. They are adsorbed directly from solution (usually of a few millimolar in alcohol) and the self-assembly process leads to a dense layer. Thiols with long alkyl chains ($n > 10$) assemble into monolayers with a two-dimensional (2D) crystalline order. Maturation and reorganization to form this perfect self-assembled layer may take several hours after the initial SAM deposition. Extensive characterizations have shown that the chains pack with a tilt angle of approximately 30° to the surface normal [FIN 96]. Thiols deprotonate upon adsorption to create strong gold–sulfur bonds:

$$R–SH + Au \rightarrow R–S–Au + e^- + H^+ \qquad [2.1]$$

These bonds show little tendency to dissociate under normal conditions. Moreover, the adsorbed molecules tend to remain in place and do not migrate around the surface.

Many different thiols with different chain lengths and functionalities are available. This is due to the widespread interest in SAM technology and the rate of development in this area is thus constantly increasing.

2.2.1.2. *Glass and similar*

Glass, silica and metal oxides provide similar types of surface bonds since their surfaces usually display a mixture of oxygen-bridged metal (or semiconductor) atoms and hydroxyl groups. The presence of these hydroxyl groups accounts for the hydrophilic behavior of such surfaces and also enables the use of organosilanes for surface modification purposes.

We have seen in the previous section that SAMs on gold are easy to prepare and even if thiols have greater uniformity than silane monolayers, gold is not always suited to biosensor applications. Indeed, most optical transducers (except for SPR sensors) rely on absorbance or fluorescence measurements: in this case, transparency of the substrate is mandatory. For such applications, glass or silica is often used in the forms of fibers or optical waveguides, and silicon dioxide is exclusively used in microfabricated structures such as field-effect transistors.

Various chemical reactions may be used to couple chloro- or alkoxysilanes to the hydroxyl groups on glass and metal oxide surfaces. Silane derivatives are available with different terminal functional groups such as amine, thiol and epoxy. These groups may be subsequently coupled to proteins or other receptors.

The most common silanization procedures involve exposing a thoroughly cleaned surface to the silane either in vapor phase or in an inert solvent such as toluene, although aqueous methods are also available using water-soluble silanes such as 3-aminopropyl-triethoxysilane (APTES) [HAL 78, KUR 93].

2.2.1.3. *Polymers*

Natural and artificial polymers are likely to become increasingly important in biosensors due to their low cost, optical properties and possibilities of processing options such as injection molding, extrusion and thermal micro-molding. Using such techniques allows producing complex shapes and surface structures (such as diffraction gratings) paving the way toward low-cost integrated devices with, for instance, combined liquid handling (and possibly on-chip sample preparation solutions) and optical components. This trend has represented a real paradigm shift in the analytical systems area with the apparition of micro total analysis systems (which may or may not contain sensing devices) [REY 02, AUR 02].

The properties of polymer surfaces intrinsically range from very hydrophobic to very hydrophilic. Most important natural polymers (carbohydrates), such as cellulose, agarose and dextran are very hydrophilic, due to the dominance of hydroxyl groups, and are thus prone to very little adsorption (favorable to minimize non-specific binding) but require the proteins to be covalently coupled to their surfaces.

On the contrary, the strong adsorption of proteins to hydrophobic polystyrene surfaces was used in the development of radioimmunoassay and enzyme-linked immunoassay [ZAN 06]. Many other polymers also adsorb strongly such as poly(L-lysine) that can then be used as an intermediary layer for surface functionalization. The large size and additive interactions of these polymeric molecules lead to almost irreversible binding under relatively mild aqueous conditions. Adsorption thus provides a convenient way to introduce functional groups such as amines onto polystyrene surfaces for subsequent covalent coupling [KAL 07]. Similar approaches can also be used with many other polymers.

2.2.2. *Entrapment methods*

Physical or chemical entrapment of bioreceptors within semi-permeable membranes or polymers is a simple and attractive approach applicable to a wide range of macromolecules and generally requiring no chemicals or procedures liable to cause denaturation or loss of viability. The main entrapment methods are briefly discussed in the following.

2.2.2.1. *Physical entrapment behind membranes*

Physical entrapment relies on the use of semi-permeable membranes, such as cellulose dialysis tubing, used to retain a solution containing the receptor molecules in a compartment adjacent to the transducer. Small molecules (in terms of molecular weight) are free to diffuse in and out of this compartment while the high molecular components are kept inside. The first demonstration of a biosensor device based on an oxygen electrode was on the basis of the physical entrapment approach [CLA 62], representing an important breakthrough in the history of the development of biosensors. The technique is still sometimes used and is particularly suitable for cells and organelles.

For large biomolecules such as enzymes and antibodies, dialysis tubing is often used as the semi-permeable membrane. The molecular weight cut-off of such membranes (determined by the size of the pores) is typically approximately 5–10 kDa for globular molecules and thus completely retains typical proteins. Other materials are also available to create membranes with various porosities, like polycarbonate and nylon.

A key limitation of this approach is the difficulty of transferring it to mass production, due to problems of sealing wet membranes over devices; thus, while it remains a useful approach for laboratory demonstrations, it is unlikely to find its way into mass-produced sensors.

2.2.2.2. *Entrapment in hydrogels*

Enzymes and antibodies can be physically entrapped within the volume of a polymer hydrogel, rather than behind a membrane, while retaining substantial activity. Polymers used for this aim can be dissolved in warm or hot water and display a gel-like consistency upon cooling (due to hydrogen bond formation). The widely used polymer is agarose, a polysaccharide obtained from marine algae. Because of its quite high porosity, agarose is suitable for whole microorganisms or organelles, rather than protein entrapment.

Polyvinyl alcohol is alternatively used in cases of proteins, e.g. enzymes, physical entrapment [REE 97]. Proteins are mixed with the polymer solution and entrapment occurs upon drying the polymer. When subsequently rehydrated, the polymer swells and the protein becomes biologically functional but cannot diffuse out of the polymer.

To reduce the amount of proteins that can leach out of an entrapping hydrogel, it may be desirable to covalently link the proteins to the polymer. To do so, it is often necessary to derivatize the protein so that it contains polymerizable functional groups that can be covalently incorporated into the growing polymer chains. The most common approach is to form acrylamide groups on the protein surface using an N-hydroxysuccinimide (NHS) reagent.

2.2.2.3. *Entrapment within conducting polymers*

Electrically conducting polymers have been used in the entrapment procedure for some biosensors based on the electrochemical transduction. The conducting properties of the polymer may either be used as part of the transduction mechanism or to perform the active grafting of biomolecules onto metallic surfaces with real-time control.

The approach most widely used relies on the pyrrole copolymerization process that is carried out under relatively mild aqueous conditions at a potential that does not damage functional bioreceptors [LIV 98]. This method that has been demonstrated with a variety of enzymes, antibodies and nucleic acids, ensures proper orientation and packing of biological receptors onto solid surfaces. With regard to the fabrication of biochips and biosensors, controlling the surface arrangement is an important feature since the sensitivity of the detection is dictated either by the amount of receptors immobilized or by their accessibility.

2.2.3. *Covalent coupling*

The attachment of biomolecules to the surface of biosensors can be achieved via covalent bonds [LIN 88]. Biomolecules such as enzymes and proteins display functional groups available for covalent immobilization onto surfaces. These include amino acid side chains (e.g. amino groups of lysine), carboxyl groups (aspartate and glutamate), sulfhydryl groups (cysteine), phenolic, thiol and imidazole groups. Covalent binding of biomolecules for biosensing applications is a favored method, and procedures resulting in minimal loss of biomolecule activity have been employed [LEG 93]. A definitive advantage of covalent coupling is that the biomolecule is generally strongly immobilized onto the surface and therefore unlikely to detach from it during use, thus this grafting method results in stable biofunctional layers. Although many covalent immobilization procedures have been developed (that involve hydroxyl groups or aldehydes, for instance), we are only describing the one of main interest for the realization of biosensors.

2.2.3.1. *Coupling to carboxylic acids*

Carboxylic acids can be readily obtained on surfaces via SAMs, oxidation or hydrolysis of polymers and polymer grafting. Since soluble biomacromolecules in most cases display an available surface amine group, immobilization to carboxylic acids is carried out via amide bond formation following the reaction shown in Figure 2.4. This is achieved by first activating the carboxylic acid using a coupling reagent. There is a large number of so-called coupling reagents, but the most popular one and the first studied is carbodiimide [VAL 09].

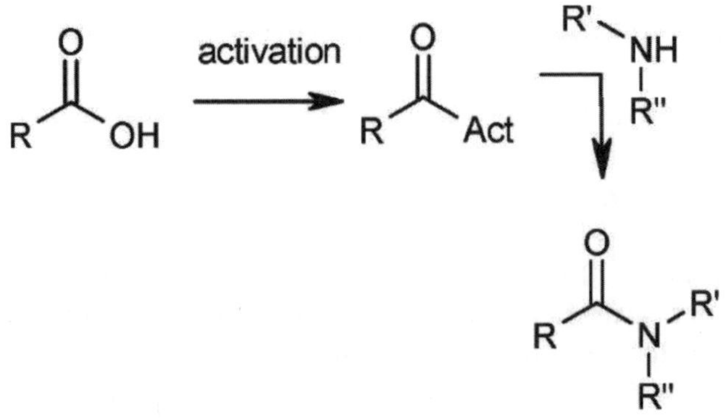

Figure 2.4. *Principle of the activation process for amide bond formation [VAL 09]*

Amide bond formation using carbodiimide involves two stages: activation by the carbodiimide reagent to form an O-acylisourea intermediate and nucleophilic displacement of the intermediate by the amine to form the final amide bond. This can be done either in organic solvents or under aqueous conditions. In the former case, the carboxylic acid is simply mixed with the amine in the presence of a mole equivalent amount of a suitable carbodiimide, but in the latter case, things are a little more complex since water-soluble carbodiimides must be used (such as 1-ethyl-3-(3-dimethylaminopropyl)carbodiimide (EDC)). To increase the efficiency of the reaction, other reagents such as NHS are used. The EDC/NHS procedure has been popularized and extensively optimized by Biacore, since it is the most common method of coupling proteins to carboxymethyl-dextran SPR chips [JOH 91].

2.2.3.2. *Reactions involving amines*

Coupling proteins and other ligands to amine surfaces has been addressed in the previous section since the same reactive groups (e.g. amine and carboxylic acid) are involved and thus the same chemistry and conditions can be used. It has to be noted that in the case of protein coupling, it is generally better to rely on the presence of carboxylic acid onto the surface to biofunctionalize, so that the activation and the coupling steps can be separated and thus protein aggregation can be avoided (which is not the case with an amine surface). However, for some small ligands such as biotin, coupling to an amine surface is very convenient since the carboxylic acid of its structure can be directly coupled using the NHS/EDC chemistry, thus avoiding the need for expensive biotin derivatives.

2.2.3.3. *Reactions involving thiols*

Thiol groups are important in antibody coupling, due to the possibility of producing Fab fragments with free thiols. They are generally coupled either by disulfide formation or by creating thioethers. Disulfides, which are relatively easily formed and stable under normal conditions, have the unique property to be readily cleaved by reduction and reformed via thiol-disulfide exchange (oxidation). This offers the possibility of regenerating the surfaces of biosensors by mild chemical stripping followed by reactivation and recoupling.

To remove a protein covalently bound to a surface, this surface is treated with a thiol-containing reducing agent such as dithiothreitol (DTT). Upon this treatment, the protein is removed, leaving free thiol groups at the surface. DTT is the reagent of choice for this step since it excises itself from the surface by an internal thiol-disulfide exchange leading to a cyclic disulfide and a free surface-bound thiol.

2.2.4. *Other capture systems*

The use of a chemical or a biochemical capture system often offers the possibility of creating generic surfaces for the specific immobilization of classes of biomolecules but usually raises issues related to the non-specific binding of other molecules onto unsaturated capture systems. The techniques presented hereafter are compatible with various options for regeneration while giving some control over the orientation of the biomolecules.

2.2.4.1. *Capture systems based on antibody-binding proteins*

The most obvious class of proteins capable of binding to antibodies is the antibodies themselves. Species and class selective antibodies, such as goat anti-human immunoglobulins (IgG), can be used to capture the antibody of interest. This type of approach is extensively used in the ELISA technology.

There are a number of proteins isolated from bacteria that have the ability to bind to the Fc region of IgGs. The best-known examples are Protein A (from *Staphylococcus aureus*) and Protein G (from Streptococcus strain G148). When immobilized onto a surface, Proteins A and G can be used as capture systems providing a favorable orientation of the anchored antibody and thus a maximized access to the binding sites [WAN 03]. A drawback of this method is that in case of random immobilization of Proteins A and G, they may not all be able to bind antibodies. Hence, although the captured antibody might function efficiently, it may be present at lower densities than if immobilized directly.

The most important advantage of using Protein A and Protein G capture systems is the ability to regenerate the capture layer, i.e. by removing the active antibody and replacing it with a different antibody under relatively mild conditions (usually by treatment with acidic buffers such as 20 mM citrate pH 3.5). These surfaces thus act as generic immunosensors platforms.

2.2.4.2. *Capture systems based on biotin-binding proteins*

Because of the extremely high affinity of avidin/biotin and streptavidin/biotin interactions (which is essentially non-reversible under normal assay conditions), the use of this capture technique is widespread in assay and labeling methodology. The various uses of the avidin/biotin capture system have been extensively reviewed in the literature [WIL 88]. Although streptavidin and avidin are interchangeable, the former is usually preferred (despite its higher cost) due to its lower non-specific binding property.

A variety of methods for immobilizing streptavidin have been used including general covalent strategies and the capture of streptavidin using covalently immobilized biotin. Streptavidin forms 2D crystalline arrays on suitable surfaces due to its "block-shaped" protein structure, with four biotin-binding sites [SPI 93]. The binding sites are distributed with two on each of nearly parallel faces of the molecule. If an appropriate surface is prepared with immobilized biotin molecules, streptavidin can bind to this surface to form a tightly anchored array, with a high density for biotin binding [MOR 92]. This is an ideal surface for the grafting of biotinylated antibodies (obtained using the NHS/EDC chemistry that couples biotin molecules randomly onto the antibody).

2.2.5. *Immobilization strategies: summary*

Despite the intense research conducted in the field of molecular grafting and surface functionalization, no generic method has yet been identified or proposed. Table 2.1 summarizes the advantages and the drawbacks offered by the various immobilization strategies discussed here. From a practical point of view, the type of bioreceptor to be immobilized and constraints imposed by the sensor and the application dictates the selection of the best-suited (or the only possible) grafting method.

Immobilization technique		Pros	Cons
Adsorption		– Simple, low cost – Adapted for single-use application	– Non-specific binding – Relatively unstable – Dependent upon pH, solvent, temperature, surface state and type of biomolecule
Entrapment	*Behind membranes*	– Simple and universal approach for macromolecules – Mild conditions – Long working life	– Difficult to mass-produce – Diffusion barrier slows response time
	In hydrogels	– Mass production potential	– Protein denaturation by free radicals
	Within conducting polymers	– May participate in the transduction mechanism – Proper orientation and packing of active bioreceptors	– Requires in-situ electro-chemical activation means
Covalent coupling		– Stable – Intimate contact with transducer – Low diffusion barrier/rapid response	– Complexity and cost of derivatization steps – Limited sites for attachment
Capture systems		– Generic surfaces – Many options for regeneration – Possibility for antibody orientation	– High-cost, complex multistep derivatization procedures – Multilayer structure may reduce signal – Non-specific binding onto components of capture system

Table 2.1. *Pros and cons of some biomolecular grafting strategies*

2.3. Conclusion

Although the variety of bioreceptors and the number of immobilization strategies might seem puzzling when addressing a practical case, the type of bioreceptor is actually dictated from the targeted application and the analyte to detect, while the actual number of techniques to be chosen is greatly reduced when taking into consideration the surface properties and chemistry of the biosensor and the biological entity to be immobilized. As a rule of thumb, the best way to move forward is to test out a few procedures to see which one gives the best compromise between speed, cost and performance for the particular system under development. What is to be kept in mind is that a pragmatic approach to address a specific application requiring biomolecules immobilization onto a solid surface is most of the time the only option.

2.4. Bibliography

[AND 95] ANDERSSON L.I., MULLER R., VLATAKIS G., *et al.*, "Mimics of the binding-sites of opioid receptors obtained by molecular imprinting of enkephalin and morphine", *Proceedings of the National Academy of Sciences of the United States of America,* vol. 92, pp. 4788–4792, May 1995.

[AND 02] ANDERSON N.L., ANDERSON N.G., "The human plasma proteome – history, character, and diagnostic prospects", *Molecular & Cellular Proteomics,* vol. 1, pp. 845–867, November 2002.

[AUR 02] AUROUX P.A., IOSSIFIDIS D., REYES D.R., *et al.*, "Micro total analysis systems. 2. Analytical standard operations and applications", *Analytical Chemistry,* vol. 74, pp. 2637–2652, June 2002.

[AYE 07a] AYELA C., VANDEVELDE F., LAGRANGE D., *et al.*, "Combining resonant piezoelectric micromembranes with molecularly imprinted polymers", *Angewandte Chemie-International Edition,* vol. 46, pp. 9271–9274, 2007.

[AYE 07b] AYELA C., ROQUET F., VALERA L., *et al.*, "Antibody-antigenic peptide interactions monitored by SPR and QCM-D – a model for SPR detection of IA-2 autoantibodies in human serum", *Biosensors & Bioelectronics,* vol. 22, pp. 3113–3119, June 2007.

[CAM 04] Campas M., Katakis I., "DNA biochip arraying, detection and amplification strategies", *Trac-Trends in Analytical Chemistry,* vol. 23, pp. 49–62, January 2004.

[CAN 04] CANZIANI G.A., KLAKAMP S., MYSZKA D.G., "Kinetic screening of antibodies from crude hybridoma samples using Biacore", *Analytical Biochemistry,* vol. 325, pp. 301–307, February 2004.

[CLA 62] CLARK L.C., LYONS C., "Electrode systems for continuous monitoring in cardiovascular surgery", *Annals of the New York Academy of Sciences,* vol. 102, pp. 29–45, 1962.

[COL 97] COLLINGS A.F., CARUSO F., "Biosensors: recent advances", *Reports on Progress in Physics,* vol. 60, pp. 1397–1445, November 1997.

[FIN 96] FINKLEA H.O., "Electrochemistry of organized monolayers of thiols and related molecules on electrodes", *Electroanalytical Chemistry: A Series of Advances,* vol. 19, pp. 109–335, 1996.

[GIZ 02] GIZELI E., LOWE C. R. (eds), *Biomolecular Sensors*, Taylor & Francis, 2002.

[GOM 00] GOMARA M.J., RIEDEMANN S., VEGA I., *et al.*, "Use of linear and multiple antigenic peptides in the immunodiagnosis of acute hepatitis A virus infection", *Journal of Immunological Methods,* vol. 234, pp. 23–34, February 2000.

[GOO 02] GOODING J.J., "Electrochemical DNA hybridization biosensors", *Electroanalysis,* vol. 14, pp. 1149–1156, September 2002.

[HAL 78] HALLER I., "Covalently attached organic monolayers on semiconductor surfaces", *Journal of the American Chemical Society,* vol. 100, pp. 8050–8055, 1978.

[HAU 98] HAUPT K., DZGOEV A., MOSBACH K., "Assay system for the herbicide 2,4-dichlorophenoxyacetic acid using a molecularly imprinted polymer as an artificial recognition element", *Analytical Chemistry,* vol. 70, pp. 628–631, February 1998.

[HEN 98] HENNION M.C., BARCELO D., "Strengths and limitations of immunoassays for effective and efficient use for pesticide analysis in water samples: a review", *Analytica Chimica Acta,* vol. 362, pp. 3–34, April 1998.

[HIL 05] HILLBERG A.L., BRAIN K.R., ALLENDER C.J., "Molecular imprinted polymer sensors: implications for therapeutics", *Advanced Drug Delivery Reviews,* vol. 57, pp. 1875–1889, December 2005.

[IVN 99] IVNITSKI D., ABDEL-HAMID I., ATANASOV P., *et al.*, "Biosensors for detection of pathogenic bacteria", *Biosensors & Bioelectronics,* vol. 14, pp. 599–624, October 1999.

[HOC 97] HOCK B., "Antibodies for immunosensors – a review", *Analytica Chimica Acta,* vol. 347, pp. 177–186, July 1997.

[JOH 91] JOHNSSON B., LOFAS S., LINDQUIST G., "Immobilization of proteins to a carboxymethyldextran-modified gold surface for biospecific interaction analysis in surface-plasmon resonance sensors", *Analytical Biochemistry,* vol. 198, pp. 268–277, November 1991.

[KAL 07] KALASHNIKOVA I.V., IVANOVA N.D., EVSEEVA T.G., *et al.*, "Study of dynamic adsorption behavior of large-size protein-bearing particles", *Journal of Chromatography A,* vol. 1144, pp. 40–47, March 2007.

[KIM 06] KIM H., COHEN R. E., HAMMOND P. T., *et al.*, "Live lymphocyte arrays for biosensing", *Advanced Functional Materials,* vol. 16, pp. 1313–1323, July 2006.

[KOC 00] KOCH S., WOLF H., DANAPEL C., *et al.*, "Optical flow-cell multichannel immunosensor for the detection of biological warfare agents", *Biosensors & Bioelectronics,* vol. 14, pp. 779–784, January 2000.

[KOS 98] KOSHKIN A. A., SINGH S. K., NIELSEN P., *et al.*, "LNA (locked nucleic acids): synthesis of the adenine, cytosine, guanine, 5-methylcytosine, thymine and uracil bicyclonucleoside monomers, oligomerisation, and unprecedented nucleic acid recognition", *Tetrahedron,* vol. 54, pp. 3607–3630, April 1998.

[KUR 93] KURTH D.G., BEIN T., "Surface-reactions on thin-layers of silane coupling agents", *Langmuir,* vol. 9, pp. 2965–2973, November 1993.

[KUS 98] KUSTERBECK A.W., CHARLES P.T., "Field demonstration of a portable flow immunosensor", *Field Analytical Chemistry and Technology,* vol. 2, pp. 341–350, 1998.

[LAS 00] LASCHI S., FRANEK M., MASCINI M., "Screen-printed electrochemical immunosensors for PCB detection", *Electroanalysis,* vol. 12, pp. 1293–1298, November 2000.

[LEE 00] LEE W.E., THOMPSON H.G., HALL J.G., *et al.*, "Rapid detection and identification of biological and chemical agents by immunoassay, gene probe assay and enzyme inhibition using a silicon-based biosensor", *Biosensors & Bioelectronics,* vol. 14, pp. 795–804, January 2000.

[LEG 93] LEGGETT G.J., ROBERTS C.J., WILLIAMS P.M., *et al.*, "Approaches to the immobilization of proteins at surfaces for analysis by scanning-unneling-microscopy", *Langmuir,* vol. 9, pp. 2356–2362, Septmber 1993.

[LIN 88] LIN J.N., HERRON J., ANDRADE J.D., BRIZGYS M., "Characterization of immobilized antibodies on silica surfaces", *IEEE Transactions on Biomedical Engineering,* vol. 35, pp. 466–471, June 1988.

[LIU 12] LIU Y., MATHARU Z., HOWLAND M.C., REVZIN A., *et al.*, "Affinity and enzyme-based biosensors: recent advances and emerging applications in cell analysis and point-of-care testing", *Analytical and Bioanalytical Chemistry,* vol. 404, pp. 1181–1196, September 2012.

[LIV 98] LIVACHE T., BAZIN H., CAILLAT P., *et al.*, "Electroconducting polymers for the construction of DNA or peptide arrays on silicon chips", *Biosensors & Bioelectronics,* vol. 13, pp. 629–634, Septmber 1998.

[MAD 93] MADOU M., TIERNEY M.J., "Required technology breakthroughs to assume widely accepted biosensors", *Applied Biochemistry and Biotechnology,* vol. 41, pp. 109–128, April–May 1993.

[MOR 92] MORGAN H., TAYLOR D.M., "A surface-plasmon resonance immunosensor based on the streptavidin biotin complex", *Biosensors & Bioelectronics,* vol. 7, pp. 405–410, 1992.

[MUL 95] MULDOON M.T., STANKER L.H., "Polymer synthesis and characterization of a molecularly imprinted sorbent assay for atrazine", *Journal of Agricultural and Food Chemistry,* vol. 43, pp. 1424–1427, June 1995.

[NIE 91] NIELSEN P.E., EGHOLM M., BERG R.H., *et al.*, "Sequence-selective recognition of DNA by strand displacement with a thymine-substituted polyamide", *Science,* vol. 254, pp. 1497–1500, December 1991.

[NUZ 83] Nuzzo R.G., Allara D.L., "Adsorption of bifunctional organic disulfides on gold surfaces", *Journal of the American Chemical Society,* vol. 105, pp. 4481–4483, 1983.

[PRI 04] PRITCHARD J., LAW K., VAKUROV A., *et al.*, "Sonochemically fabricated enzyme microelectrode arrays for the environmental monitoring of pesticides", *Biosensors & Bioelectronics,* vol. 20, pp. 765–772, November 2004.

[RAM 97] RAMSTROM O., YE L., "Molecularly imprinted materials – their use in separations, immunoassay-type analyses and syntheses", *Abstracts of Papers of the American Chemical Society,* vol. 213, p. 129-IEC, April 1997.

[REE 97] REETZ M.T., "Entrapment of biocatalysts in hydrophobic sol-gel materials for use in organic chemistry", *Advanced Materials,* vol. 9, p. 943, October 1997.

[REY 02] REYES D.R., IOSSIFIDIS D., AUROUX P.A., *et al.*, "Micro total analysis systems. 1. Introduction, theory, and technology", *Analytical Chemistry,* vol. 74, pp. 2623–2636, June 2002.

[ROG 98] ROGERS K.R., MASCINI M., "Biosensors for field analytical monitoring", *Field Analytical Chemistry and Technology,* vol. 2, pp. 317–331, 1998.

[SAV 10] SAVORY N., ABE K., SODE K., *et al.*, "Selection of DNA aptamer against prostate specific antigen using a genetic algorithm and application to sensing", *Biosensors & Bioelectronics,* vol. 26, pp. 1386–1391, December 2010.

[SHA 13] SHALEV G., LANDMAN G., AMIT I., *et al.*, "Specific and label-free femtomolar biomarker detection with an electrostatically formed nanowire biosensor", *Npg Asia Materials,* vol. 5, p. 7, March 2013.

[SPI 93] SPINKE J., LILEY M., GUDER H.J., *et al.*, "Molecular recognition at self-assembled monolayers – the construction of multicomponent multilayers", *Langmuir,* vol. 9, pp. 1821–1825, July 1993.

[STE 90] STEWART D.J., PURVIS D.R., LOWE C.R., "Affinity-chromatography on novel perfluorocarbon supports – immobilization of CI reactive blue-2 on a polyvinyl alcohol-coated perfluoropolymer support and its application in affinity-chromatography", *Journal of Chromatography,* vol. 510, pp. 177–187, June 1990.

[STE 00] STEFAN R.I., VAN STADEN J.F., ABOUL-ENEIN H.Y., "Immunosensors in clinical analysis", *Fresenius Journal of Analytical Chemistry,* vol. 366, pp. 659–668, March–April 2000.

[SUR 01] SURUGIU I., SVITEL J., YE L., *et al.*, "Development of a flow injection capillary chemiluminescent ELISA using an imprinted polymer instead of the antibody", *Analytical Chemistry,* vol. 73, pp. 4388–4392, Septmber 2001.

[SUS 03] SUSMEL S., GUILBAULT G.G., O'SULLIVAN C.K., "Demonstration of labeless detection of food pathogens using electrochemical redox probe and screen printed gold electrodes", *Biosensors & Bioelectronics,* vol. 18, pp. 881–889, July 2003.

[TOT 01] TOTHILL I.E., "Biosensors developments and potential applications in the agricultural diagnosis sector", *Computers and Electronics in Agriculture,* vol. 30, pp. 205–218, February 2001.

[VAL 09] VALEUR E., BRADLEY M., "Amide bond formation: beyond the myth of coupling reagents", *Chemical Society Reviews,* vol. 38, pp. 606–631, 2009.

[VLA 93] VLATAKIS G., ANDERSSON L.I., MULLER R., *et al.*, "Drug assay using antibody mimics made by molecular imprinting", *Nature,* vol. 361, pp. 645–647, February 1993.

[VOD 00] VO–DINH T., CULLUM B., "Biosensors and biochips: advances in biological and medical diagnostics", *Fresenius Journal of Analytical Chemistry,* vol. 366, pp. 540–551, March–April 2000.

[WAH 91] WAHLGREN M., ARNEBRANT T., "Protein adsorption to solid-surfaces", *Trends in Biotechnology,* vol. 9, pp. 201–208, June 1991.

[WAN 97] WANG J., RIVAS G., CAI X., *et al.*, "DNA electrochemical biosensors for environmental monitoring. A review", *Analytica Chimica Acta,* vol. 347, pp. 1–8, July 1997.

[WAN 00] WANG J., *Analytical Electrochemistry*, 2nd ed., 2000.

[WAN 03] WANG Z.H., JIN G., "Feasibility of protein A for the oriented immobilization of immunoglobulin on silicon surface for a biosensor with imaging ellipsometry", *Journal of Biochemical and Biophysical Methods,* vol. 57, pp. 203–211, September 2003.

[WIL 88] WILCHEK M., BAYER E.A., "The avidin biotin complex in bioanalytical applications", *Analytical Biochemistry,* vol. 171, pp. 1–32, May 1988.

[YAN 97a] YANG M.S., MCGOVERN M.E., THOMPSON M., "Genosensor technology and the detection of interfacial nucleic acid chemistry", *Analytica Chimica Acta,* vol. 346, pp. 259–275, July 1997.

[YAN 97b] YANG Z.H., YU H., "Preserving a globular protein shape on glass slides: a self-assembled monolayer approach", *Advanced Materials,* vol. 9, p. 426, April 1997.

[YE 04] YE L., HAUPT K., "Molecularly imprinted polymers as antibody and receptor mimics for assays, sensors and drug discovery", *Analytical and Bioanalytical Chemistry,* vol. 378, pp. 1887–1897, April 2004.

[ZAN 06] ZANGAR R.C., DALY D.S., WHITE A.M., "ELISA microarray technology as a high-throughput system for cancer biomarker validation", *Expert Review of Proteomics,* vol. 3, pp. 37–44, February 2006.

3

Patterning Techniques for the Biofunctionalization of MEMS

The biofunctionalization of miniaturized biosensors requires the localized surface immobilization of bioreceptors through one of the dedicated grafting strategies discussed in Chapter 2. As biosensors are getting smaller, coating their surfaces with bioactive layers is becoming challenging. In many cases, commonly used immersion is not appropriate because selective functionalization of active areas can be difficult to achieve. Besides, in order to fully take advantage of the opportunity to create arrays of sensors by using large-scale microfabrication techniques, the spatial control of biofunctionalization is not only mandatory but should also provide solutions for the grafting of different types of bioreceptors onto the different elements of the array. We should also bear in mind that ultimately the non-reactive areas of the chip containing multiple sensors functionalized with probe molecules should be adequately coated with an antifouling film (e.g. with a protein or polymer layer) in order to lower the probability of adsorption of target molecules anywhere other than on sensitive areas and hence to permit their detection at ultra-low concentrations. The choice of the passivation strategy (in terms of type of antifouling agent and grafting chemistry) should be addressed globally within the biofunctionalization task. Finally, another requisite for biofunctionalization tools is the ability to handle biomolecules without causing their denaturation. Given these facts, a number of approaches are available, including those relying on the use of micromachined pipettes, inkjet or elastomeric stamps, and the choice of the dedicated technique must be carried out in accordance with the sensor specifications and the type of bioreceptor.

3.1. What is surface patterning?

Patterning techniques are the methods used to spatially control the deposition/location of materials at the surface of a substrate. When combined with one of the immobilization strategies presented in Chapter 2, these techniques enable the grafting of biomolecular probes solely at the active surface of a biosensor with, if possible, means to graft various probes onto a biosensor array for multiplexing purpose.

In this chapter, we present the different classes of patterning techniques that are currently being used for substrate biofunctionalization. More precisely, we describe these techniques that can be organized in two categories – the direct writing (where patterns are created from scratch) and the replication of patterns (where patterns are duplicated from prefabricated masks) – with the following characteristics used as comparative elements in mind:

– minimum resolution and feature size;

– uniformity and reproducibility of the patterns;

– quality and functionality of the deposited materials;

– constraints related to the substrate state;

– pattern alignment and multiplexing capabilities.

An important number of tools described in this section rely on the delivery of liquid droplets onto a substrate because these tools are particularly suitable for handling biomolecules without causing their denaturation. In fact, biofunctionalization tools based on droplet spotters are currently used to fabricate the majority of microarrays used in clinical laboratories.

3.2. Direct biopatterning in liquid phase

Ink-based direct writing techniques rely on the local delivery of materials onto a surface via a liquid meniscus or a droplet. Various methods exist for spatial delivery of liquid samples, the most appropriate method depends on the scale of the array to be produced, the minimum feature size and the throughput required. The range of feature sizes that can be achieved spans an impressive 12 orders of magnitude in delivered volumes from a microliter (10^{-6} l) to attoliter (10^{-18} l). Figure 3.1 compares the dimensions and volumes involved in liquid phase surface micro- and nanopatterning and positions them with respect to the intuitive techniques such as pipetting or inkjet printing.

Two different classes of methods are generally used in droplet-based patterning: the non-contact and the contact method. While the former consists of projecting liquid drops on the surface from a droplet ejector, the latter consists of transferring droplets due to the liquid meniscus formed between the depositing tool (usually, a tip or a pen) and the substrate.

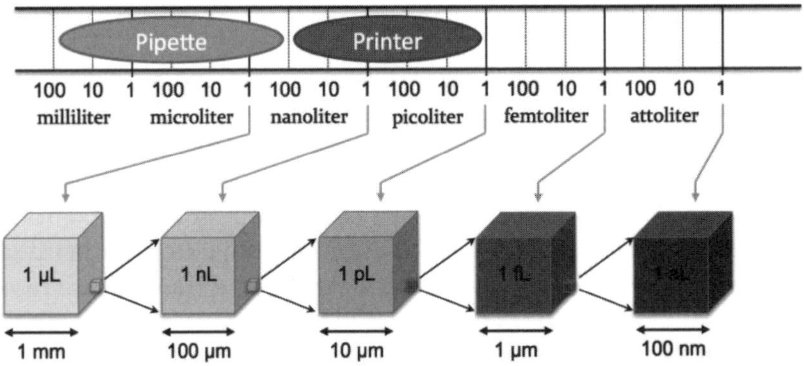

Figure 3.1. *Volumetric scale providing the dimensions involved in the description of small volumes. As a reference, the range of volumes of liquid handled by two of the most common drop dispensers is shown on the scale*

3.2.1. *Ink delivery by non-contact methods*

Non-contact liquid spotting methods consist of ejecting a droplet from a nozzle placed above the surface. They offer the advantage of being compatible with any type of substrate, especially those with irregular surfaces that are difficult to pattern with other tools, but the alignment with existing patterns can be quite challenging.

3.2.1.1. *Inkjet printing*

Inkjet printing is a mature technology that is used in most of the commercial and professional computer printers. It is fairly inexpensive, pretty fast and many printers can produce high-quality outputs. The inkjet underlying principle is based on the ejection of liquid droplets from a chamber through a nozzle. In drop-on-demand systems, there are usually two means of expelling the drop. Either a heater is used to create a bubble in the closed chamber or a piezoactuator is used to compress the chamber until the liquid is propelled out of the nozzle. Piezoelectric inkjet printing is usually more reliable in terms of jetting precisions and controlled volumes [REI 05]. It allows printing a wider range of inks, as thermal inkjet relies solely on water-based inks.

Inkjet has lately been extensively used for surface patterning because of its low-cost, ease-of-use and flexibility regarding the variety of materials that can be deposited. Inkjet printing is especially very promising for the deposition of biological compounds. DNA and proteins can be printed without alteration for the fabrication of functional bioarrays [GOL 00, HUG 01]. Xu *et al.* [XU 05] were able to print mammalian cells and showed that 90% of the deposited cells were not damaged during the nozzle firing. In fact, the inkjetting technology has actually been proven to be an excellent choice for the functionalization of fragile MEMS structures, such as cantilever-based biosensors (Figure 3.2) [BIE 04a].

In addition to its proven versatility, parallelization of inkjet is straightforward and can benefit from mass-fabrication techniques resulting in low-cost devices [SHE 06]. However, the drawback of inkjet technology mainly concerns the requirement of inks with specific properties in terms of viscosity and surface tension. Inkjet is also very sensitive to nozzle clogging, due to the solvent drying or particle aggregation, and to drop dripping. The uniformity of the printed features is also a big concern because the formation of pinholes can easily occur. However, Bietsch *et al.* [BIE 04b] have recently proved that high-quality self-assembled monolayers (SAMs) could be patterned. Finally, the constraints arising from surface tension and viscous flow through a small hole seem to make unlikely the direct printing of features much smaller than 10 µm [CAL 01].

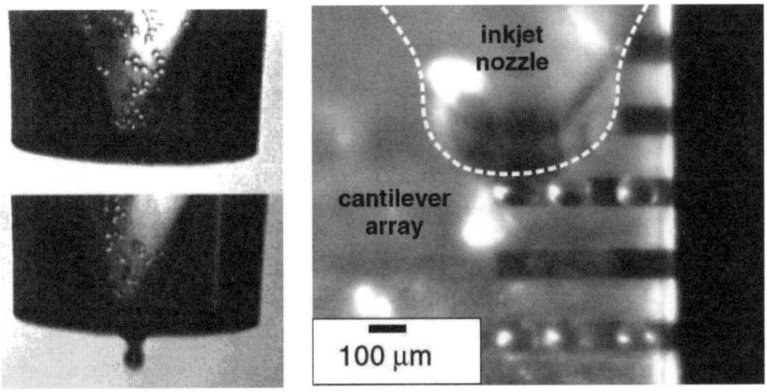

a) b)

Figure 3.2. *a) Human liver cells being jetted in a phosphate buffer. b) Inkjet printing of individual droplets onto a cantilever array*

3.2.1.2. *Electrospray and E-jet printing*

In the electrospray type of droplet generation, the stream of liquid coming out of a capillary is caused to split into numerous droplets by applying a direct current (DC) voltage. The generated electric field causes the formation of a Taylor cone, leading to a spray of fine ionized droplets. Electrospray ionization was first applied in the 1980s for mass spectrometry using a stainless steel needle [YAM 84]. However, Yogi *et al.* [YOG 01] have used a similar method to create on-demand droplets and realize a microarray of rhodamine B (rhB) molecules. This technique yields to the routine production of pico- to femtoliter droplets from a glass capillary by using 1,000 V and 10 ms pulse voltage (Figure 3.3). Recently, the Rogers research group has made a low cost, automatic <10 µm resolution desktop system for electrohydrodynamic jet (E-jet) printing [BAR 10]. The E-jet printing technique based on electrospray has been found to be suitable for the preparation of DNA arrays. The electrospray technique is thus very attractive for depositing minute amounts of liquid; however, it requires complicated apparatuses and the generation of high voltages to operate.

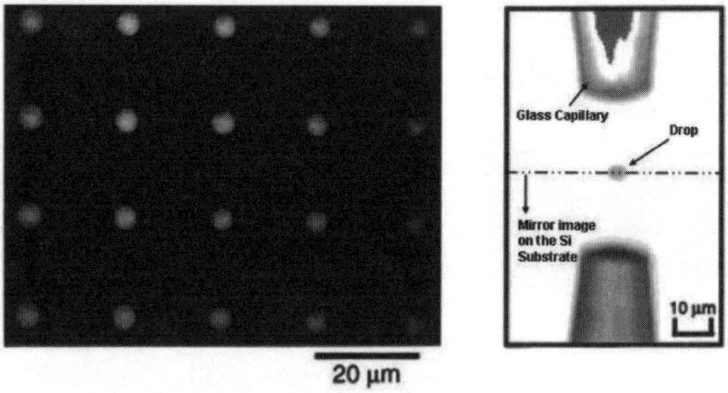

a) b)

Figure 3.3. *a) Microarray of rhB molecules deposited by the pulsed-filed spotter and b)side-view video image of a droplet of rhB solution in water after deposition [YOG 01]*

3.2.1.3. *Capillary-free jetting mechanism*

The capillary-free jetting technique, proposed by Barron *et al.*, is based on the use of a focused laser pulse to transfer material from a target support onto a substrate [BAR 05]. The laser hits the back of the target at a location of a specific microwell, inducing the heating-up of the material contained in the well via a laser absorption layer, and leads to the formation of a jet. The jet breaks up into a droplet that is projected onto the substrate. The spot diameter can be tuned by adjusting the laser fluence. This technique has been successfully used to fabricate bovine serum albumin (BSA) protein arrays. Immunoassays with anti-BSA have been carried out, demonstrating the immunoreactivity of the proteins. This result shows that the energy transfer induces little heating and demonstrates that the capillary-free jetting has the potential to be very well suited for protein patterning.

3.2.2. *Ink delivery by contact methods*

In contact methods, a solid tip or the extremity of a capillary touches the surface to pattern, allowing species to be transferred or liquid drops to be deposited. The advantages of having a direct contact between the tool and the substrate are numerous. First, the contact method is a passive delivering technique because the transfer occurs via the capillary forces acting on the liquid. Thus, there is no need for liquid actuation means and tools are greatly simplified. Moreover, sensors can be incorporated to the tool itself to have a direct control of the deposition process and direct information on the tool position. However, this method depends greatly on the physical and chemical states of the surface to pattern.

3.2.2.1. *Micro- and nanopipettes*

Pipettes can be used in contact mode for the local deposition of liquids. Any capillary in contact with a surface, providing the substrate is not too hydrophobic, leads to the formation of a drop. Downscaling of pipetting techniques was provided by a very interesting patterning solution relying on the use of a double-barrel nanopipette, fabricated from a glass capillary [ROD 05]. The main strength of this method is the ability to independently deliver two different species from each barrel of a single tip (Figure 3.4). The use of electrodes placed inside each barrel enables the selection of the depositing barrel and the voltage control of deposition parameters such as the distance between the tip and the surface and the volume of liquid delivered.

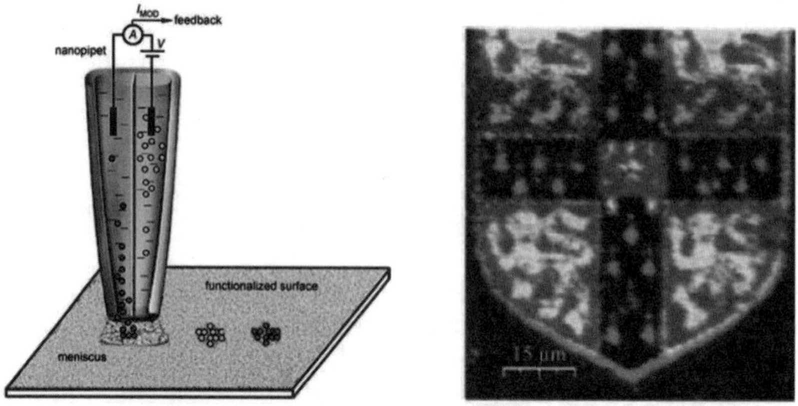

a) b)

Figure 3.4. *a) Principle of double-barreled pipette deposition. b) Double-barreled pipette-printed reproduction of the University of Cambridge crest. The image was produced with two DNA species, one labeled with Alexa 647 and the other with rhodamine green [ROD 05]*

Microfabricated tools that mimic the capillary pipettes have also been introduced lately [GUE 05]. They consist of silicon or polymer tubes arranged on a surface. Liquid is provided on one side of the device and flows into the tubes to reach the depositing surface. The use of microfabrication technologies allows us to fabricate the inner and outer diameters of tube with high precision and small dimensions, while pushing the parallelization of this deposition method. However, these tubes are usually not flexible and cannot undergo large stresses and strains, thus microfabricated pipette arrays need to be operated very carefully. Alignment with prefabricated patterns can also be a non-trivial issue.

3.2.2.2. *In-plane microcantilevers for liquid delivery*

Recently, direct liquid-delivery techniques inspired by the stainless steel pins currently used to fabricate biochips have been proposed. Commercial systems (arrayers from ArrayIt®) based on quills and pins allow printing spots with minimum diameters of 60 µm. The needles are machined in a serial manner and are usually very expensive. Moreover, because the pins are made of stainless steel and the deposition occurs by a direct contact of the pin tip with the surface, the tips are easily damaged, resulting in inconsistent drop sizes. Finally, parallelization can only be obtained by manually placing several needles on the same support (Figure 3.5).

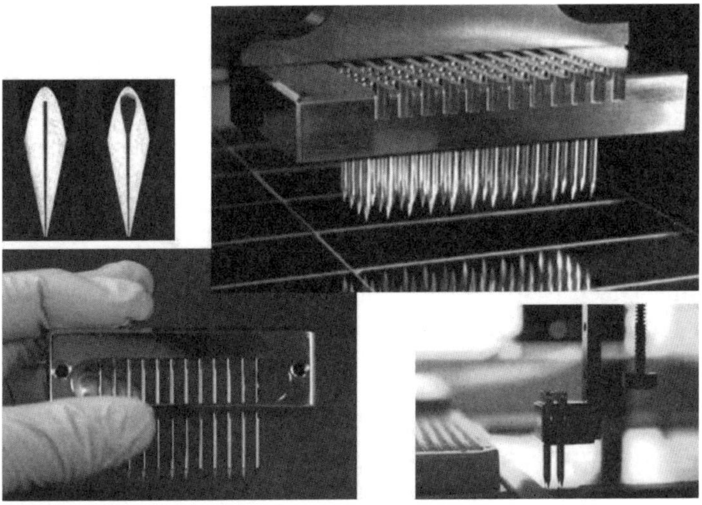

Figure 3.5. *Photographs of a commercial spotting system for biochip fabrication (ArrayIt®, TeleChem International, Inc.)*

The idea of using silicon microcantilevers as printing pins emerged in the early 2000s simultaneously in Europe [BEL 03] and in the United States [TSA 03, XU 04], leading to similar devices and commercial equipment, Nano eNabler™ from BioForce Nanosciences, Inc. This idea offers several advantages: arrays of cantilevers can be incorporated onto the same chip by standard microfabrication techniques, resulting in cheap and disposable devices. Because of the micrometric precision of photolithography, these cantilevers can be made very small, enabling the deposition of tiny droplets. Droplet size is controlled by the contact time between the tip and the substrate and drop diameters typically range from micrometers to tens of micrometers. Since the tip used for liquid transfer is in the plane of the cantilever, it can be fed directly by a fluidic system placed on top or inside the cantilever. Usually the fluidic system consists of an open channel etched in the cantilever, thus solving any clogging issue (however this usually requires us to work with low evaporation rate solvents, e.g. glycerol). The size of the channel is usually large enough to allow printing hundreds of spots without refilling the cantilever that can be then loaded with an external chip [BER 08]. However, a liquid reservoir can also be incorporated on the chip bearing the cantilevers.

To prove the importance of this technique for the increase of density of biochip microarrays, proofs of concepts were demonstrated by directly depositing DNA and proteins drops onto previously functionalized surfaces. Micromachined membranes were also biofunctionalized with molecular imprinted polymers (MIPs) for herbicide detection application using this spotting method [AYE 07]. Recently, a very interesting approach, derived from the electrospotting technique, was demonstrated to create robust and long-lasting bioarrays through the electrodeposition of biofunctionalized polypyrroles [DES 07]. These silicon cantilever pens are thus very attractive for biofunctionalization purposes because of their ability to deliver sub-picoliter droplets of biological material in a high-throughput and controlled manner (integrated sensing schemes can be implemented for automated spotting [LEI 08]), and because of their simplicity of operation and low cost.

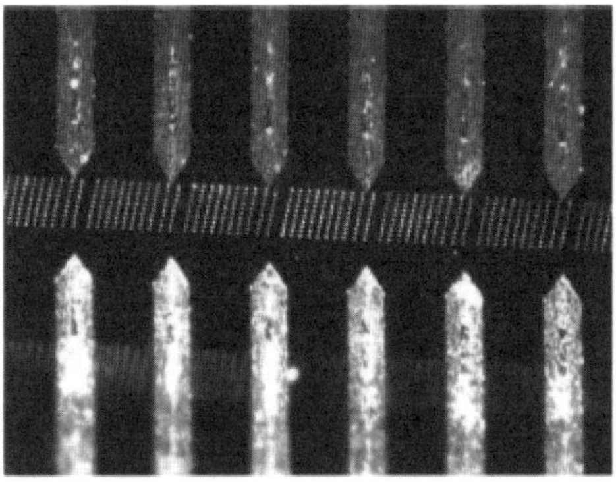

a)

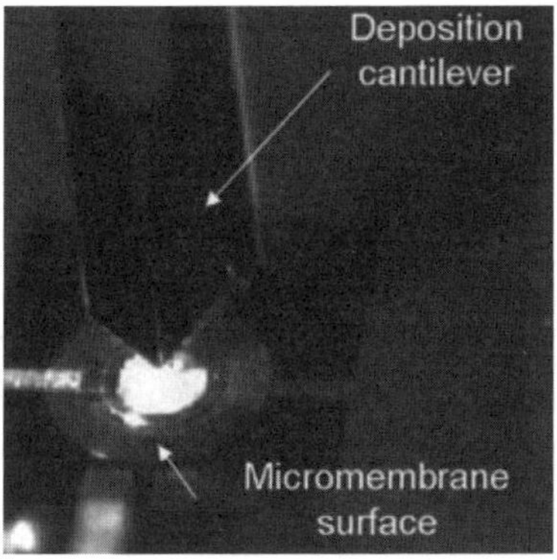

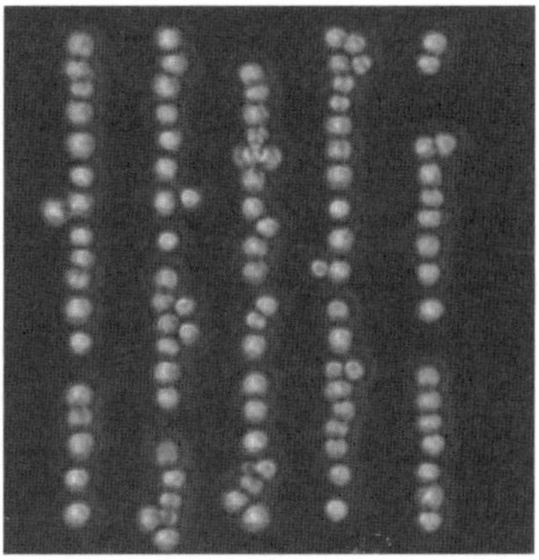

b) c)

Figure 3.6. *The silicon microcantilever-based spotting system, Bioplume. a) optical picture of the Bioplume cantilever array after depositing a droplet matrix (each cantilever is 120 µm large and 1,500 µm long) [LEI 08]. b) Optical picture of a cantilever loaded with MIP precursor solution during the functionalization of a MEMS resonant membrane [AYE 07]. c) Microscope images of bound lymphocytes onto a microarray of anti-IAb antibody spots fabricated with Bioplume (the anti-IAb lines are 7 µm in size with a 20 µm pitch) [ROU 09]*

3.2.2.3. *Scanning probe lithography-based liquid delivery systems*

Scanning probe techniques are based on atomic force microscopy (AFM) for which silicon cantilevers with sharp tips scan a substrate to create an image of the surface topography with nanometric precision. The original idea of the scanning probe lithography (SPL) technique is to use the cantilever tip to deposit chemical and biological materials. The interesting feature of SPL is that the same tip can be used to write patterns and to image them. SPL methods take advantage of the high resolution of AFMs, so far resulting in sub-100 nm minimum feature sizes. This high resolution is balanced by the long writing times necessary to pattern microareas.

Butt *et al.* have first demonstrated the use of a scanning force microscope (SFM) tip to deposit organic material [JAS 95]. In their experiments, an SFM tip was covered with octadecanethiol (ODT), by dipping it into liquid ODT, and then brought into contact with a mica surface. Upon contact with the surface, ODT was transferred from the tip onto the surface. This technique has been further developed (and popularized) by Mirkin under the name dip-pen nanolithography (DPN) [GIN 04]. In this technique, the ink loaded onto the AFM tip is delivered to the substrate via a water meniscus that forms between the tip and the surface (Figure 3.7). This meniscus serves as a medium for ink transport with nanometric precision and is thus fundamental for the success of ink delivery. This requires working in a suitable humid environment with enough water vapor to allow proper condensation on the tip. Importance of controlling the humidity level is necessary for reproducible patterning results [SUI 05]. DPN was initially used to pattern alkylthiols (e.g. MHA and ODT) onto gold films with an overwriting capability and a 15 nm resolution [PIN 99, HON 99]. Its writing potential has rapidly been extended to numerous SAMs and surfaces (e.g. hexamethyldisilazane (HMDS) on semiconductor surfaces [IVA 01]), allowing the indirect patterning of bigger entities. The indirect method consists of first patterning the surface with functional chemistry via DPN to generate a template for the subsidiary grafting of DNA strands [DEM 01, ZHA 02] or proteins

[ZHA 03, LEE 04]. Since only the template fabrication is carried out locally (unlike the immobilization procedure), it is difficult to produce patterns of different compounds on the same surface. For this purpose, the direct patterning of large molecules has been demonstrated. DNA and proteins, adsorbed on tips, have been deposited without altering their biofunctionality [LIM 03, LEE 03, CHU 05].

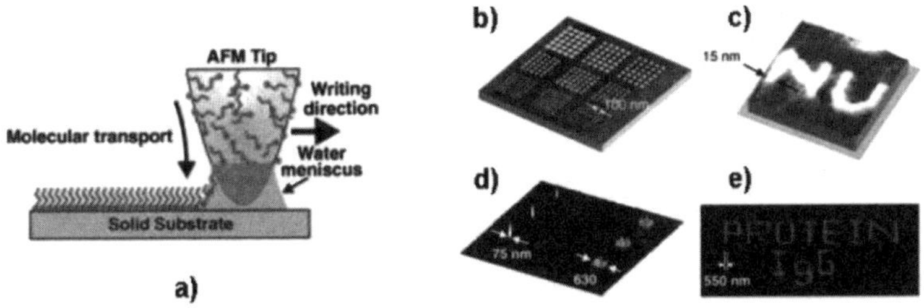

Figure 3.7. *Dip-pen lithography, principle of operation and fabricated patterns [Gin 04]. a) Schematic representation of the DPN process. b) Nanoscale dot array written on a polycrystalline Au surface with MHA. c) Nanoscale letters written on an Au (111) surface with MHA with DPN. d) AFM image of 25 and 13 nm gold nanoparticles hybridized to surface DNA templates generated with direct-write DPN. e) Fluorescence image of direct-write DPN patterns of fluorescently labeled immunoglobulin G (IgG) on SiOx*

Because AFM tips are fabricated with standard micromachining processes, the fabrication of numerous DPN on the same chip is easily achievable [SAL 05] and force sensors can be integrated to the cantilevers. The possibility to address each cantilever of an array is an essential characteristic for producing arbitrary features and for controlling the writing process.

In scanning probe contact printing (SPCP), the cantilever tip is made of an elastomeric material [WAN 03, ZHA 04]. The tip that is usually made of polydimethylsiloxane (PDMS) is used to print chemicals analogously to a microcontact stamp. A chemical ink is first absorbed into the PDMS tip and each contact of the tip with the surface leads to the transfer of molecules onto the substrate in the form of a pixel print. Arbitrary patterns can thus be formed in dot-matrix manner. Unlike DPN, SPCP can be operated at ambient conditions and features are inherently larger. Because both DPN and SPCP techniques rely on the use of AFM probes, they can all be integrated onto the same chip, including printing and imaging capabilities. Wang *et al.* have demonstrated the fabrication of such an integrated device (Figure 3.8), with different probes being able to perform a dedicated function [WAN 05]. This multifunctional probe array is, therefore, capable of performing a rich variety of operations with minimal chemical cross talk and high registration accuracy. However, in both DPN and SPCP methods, patterning is a slow process where writing steps must often be alternated with necessary reloading procedures because of the tiny amount of ink material that a probe can carry.

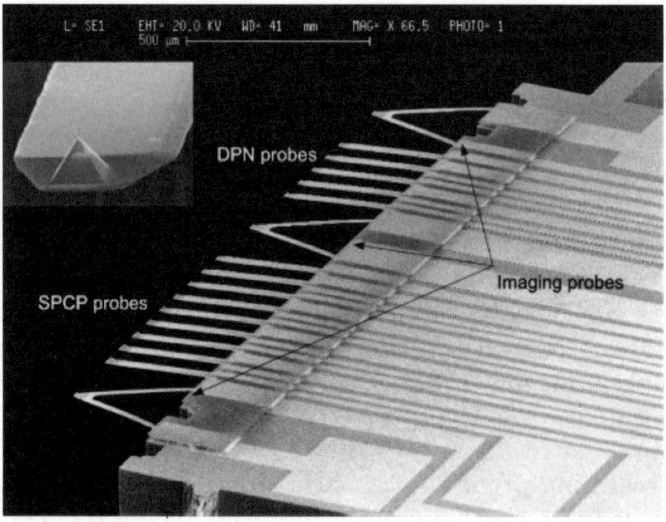

Figure 3.8. *SEM micrograph of a multifunctional probe array including five DPN probes, nine SPCP probes and three imaging probes [WAN 05]*

In order to extend the printing capability of AFM-based techniques, a fluidic capillary and a liquid reservoir can be added to the cantilever tip. This trick provides the main advantage for solving the issue of the rapid ink depletion observed with dip-pen and SPCP. For this purpose, Lewis *et al.* have proposed a nanofountain pen system based on the use of cantilevered quartz micropipettes mounted onto the probe of an AFM [LEW 99]. The local printing of proteins has been carried out with this system [THA 03]. However, despite its interesting capability, this system is not suitable for parallelization and integration. Recently, a very promising approach has been proposed by Meister *et al.* to overcome these drawbacks. The key feature of their nanodispensing system (NADIS) is the deposition of liquids through apertured scanning force microscopy probe tips. The first probes were fabricated using a dedicated process, allowing the creation of an open reservoir on top of the cantilever with a nanochannel at the apex [MEI 03]. An alternative fabrication method, which relies on the modification of commercial AFM tips by FIB milling, has also been proposed [MEI 04]. Nanodrops of 20 nm polystyrene particles were deposited with this system. The open reservoir on top of the cantilever raises two important issues. The loading of the cantilever is very delicate, and given the small volume of the drop reservoir only liquids with a low evaporation rate can be used as solvent mediums. Hence, Deladi *et al.* have recently fabricated a cantilever with a close channel, allowing a continuous liquid supply to the cantilever tip [DEL 04]. Kim *et al.* presented a similar device that was used to pattern 40 nm wide lines of MHA on gold substrates [KIM 05] (Figure 3.9). However, these closed-channel configuration tools can experience clogging, thus leading to a failure to write. Besides, simple cleaning procedures are difficult to foresee.

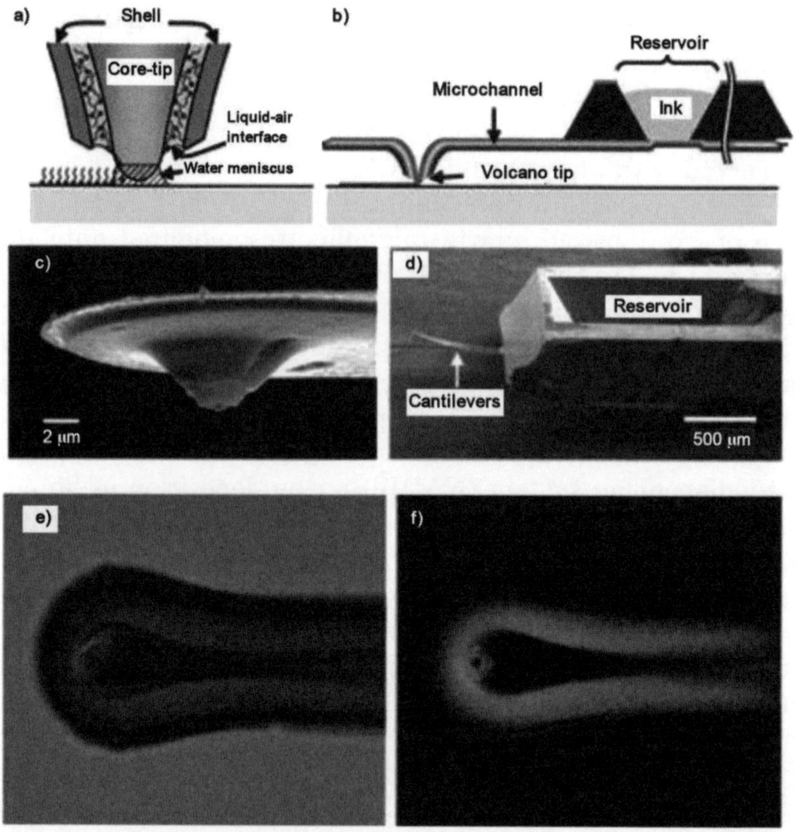

Figure 3.9. *Nanofountain probe (NFP) [KIM 05]. a) and b) Writing mechanism of the NFP device. c) and d) SEM images of a volcano dispensing tip and an on-chip reservoir. Flow test with a fluorescent solution: e) transmission optical photograph with empty channel and f) transmission fluorescence photograph showing the liquid in the fluidic channel*

3.3. Replication of patterns

Replication consists of duplicating patterns from a mask, a master or a stamp onto a substrate with high fidelity and reliability. These methods enable the fabrication of multiple copies rapidly and, thus, are well suited for mass fabrication.

3.3.1. *Photolithography*

Photolithography has been developed into a highly reproducible and high-resolution (submicrometer) patterning technique by the semiconductor industry, enabling subsequent material etching or deposition.

Photoresists can be used to protect specific areas of a substrate while others are derivatized by an appropriate surface modification procedure. Hydrophilic/hydrophobic patterns have, for instance, been produced this way for directing the growth of neuronal cell cultures onto surfaces [CLA 93]. Alternatively, the photoresist can be used after chemically modifying the entire surface to protect areas where biomolecules are to be subsequently grafted, while the reactive chemistry of non-protected areas is destroyed by, e.g., exposure to oxygen plasma [LEI 12].

The main drawback of using photolithography for biofunctionalization purposes is that current photoresists are not compatible with aqueous conditions. Besides, the organic solvents, the high temperature and radiation intensity used to pattern photoresists are sources of protein denaturation; therefore, the patterning step must be completed before the biomolecules are introduced. Nevertheless, lithography was adapted in the late 1990s for the manufacture of DNA chips [SCH 98]. In this method, repetition of lithographic steps with different chemical reagents and appropriate masking enables the construction of oligonucleotide chains with different DNA bases in desired sequences, although it is suitable only when a small number of sequences are involved.

3.3.2. *Light-induced patterning strategies*

Biofunctional patterns can be created using direct photochemical immobilization methods, through photocoupling and photodeprotection reactions. Indeed, a number of chemical functional groups, such as aryl azides, aryl diazirines and benzophenones, transform into highly reactive intermediates (nitrene, carbine and O-radical, respectively) when activated by light in the ultraviolet (UV) range [GIZ 02] , which can subsequently react with proteins and other biomolecules.

The general procedure for using any of the above reagents is first to immobilize the reagent onto the surface to functionalize via a suitable coupling method (a variety of amine- and thiol-reactive derivatives with aryl azide or benzophenone groups are commercially available). Then, the surface is exposed to UV light through a mask for selective radiation for a few minutes to achieve the coupling [BAR 98]. Arrays can be obtained by repeating the coupling cycle with different biomolecules and masks, and functionalization can be achieved within enclosed microchannels [BAL 05].

The alternative to directing photoactivation and coupling is to use a photodeprotection strategy to "unmask" a functional group and make it available for a conventional covalent coupling reaction. Methods for photodeprotecting amine groups are particularly well developed and the amines can subsequently be used in a wide range of covalent coupling methods. This strategy was used in the development of a photoactivatable polyacrylamide gel containing protected amino groups that could be photoexposed and used for coupling of antibodies and antigens [SAN 98].

3.3.3. *Microcontact printing*

Microcontact printing (μCP), an alternative biofunctionalization technique that was developed in the 1990s, relies on the use of an elastomeric stamp to transfer ink onto a substrate in a very simple and convenient way [MAR 95]. Soft stamps usually made up of polydimethylsiloxane (PDMS) exhibit grooves patterned onto a flat surface. The stamp is first soaked in the solution of ink containing the species of interest and then dried. It is then brought into contact with the surface of a substrate to allow the transfer of species from the surface of the stamp onto the substrate before being released. Because the material inside the groove does not come into contact with the substrate, it is not transferred, resulting in the replica of the stamp pattern onto the substrate.

Microcontact printing has been demonstrated for the patterning of SAMs and it works well for other chemical species, such as catalysts, dendrimers and organic reactants. This technique has been used to modify the surface chemistry of substrates with the aim, for example, of controlling the growth of cell cultures [SIN 94] or inducing controlled chemical gradients [KRA 05]. Elastomeric stamps provide an ideal interface between biological and non-biological entities. Thus, it is especially suitable for the deposition of lipid bilayers, proteins (Figure 3.10) and DNA [BER 00, LAN 04]. Soft lithography is a promising tool for the patterning and biochemical functionalization of large areas at a very low cost. Apart from its simplicity (relevant results may be achieved in laboratory without requiring complex equipment and clean-room facilities), microcontact printing is suitable for the functionalization of curved surfaces and other three-dimensional surfaces, provided there is access for a shaped rubber stamp. This considerably increases the impact of this technique.

3.3.4. *In-flux functionalization*

Pattern replication and surface functionalization can also be conducted using a microfluidic device [JUN 01]. Basically, an open channel fluidic network is placed on the surface to pattern so that the liquid flowing through the channels comes in contact with the surface. The liquid serves as a transport and reaction medium that can be used to deposit cells or biomolecules from a surface by adsorption or specific recognition. It has recently been demonstrated that when combined with another patterning technique, e.g. microcontact printing, microfluidic patterning is a very powerful method for creating multiplexed and localized protein patterns within microchannels [DID 12].

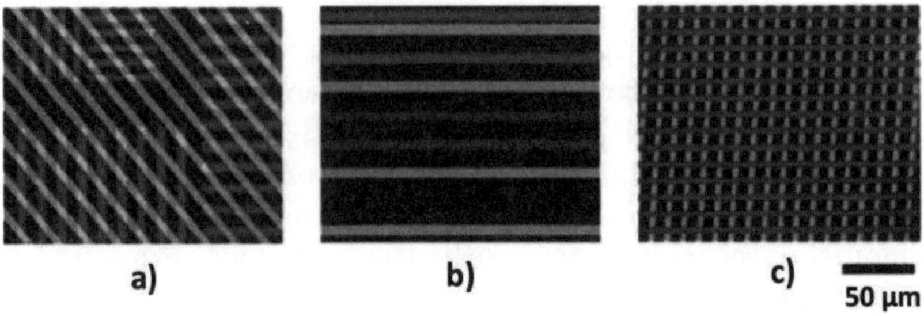

Figure 3.10. *Fluorescent images of microcontact printed proteins onto a substrate [BER 00]. a) Two different proteins subsequently printed using different PDMS stamps. b) Sixteen different proteins (some of them without a fluorescent label) patterned onto the polystyrene surface of a cell culture dish using a stamp inked by means of a fluidic network. c) Three different proteins simultaneously printed onto a glass slide by a flat stamp. The stamp was inked homogeneously with the first protein then brought into contact with a patterned silicon surface. In the regions of contact, the proteins were transferred to the silicon (lift-off), opening up areas for subsequent protein adsorption onto the stamp. This procedure can be repeated using objects made up of different materials such as silicon, glass or plastic to lift-off patterns of proteins from the stamp*

3.4. Conclusions

The various techniques introduced in this chapter enable one to pattern a wide range of biological materials with feature characteristics specific to each method. According to the requirements of a targeted application, some tools are better than others. In fact, the selection of a suitable tool is governed by the surface state of the sensor, the type of bioreceptor and the selected grafting method. The production scale and the financial means are also important parameters to be taken into account when choosing a tool. For industrial production, which requires the functionalization of a large number of similar biosensors with high yields, the use of replication techniques is definitely more appropriate. Direct-writing techniques, which enable the straightforward deposition of small amounts of biological matter in a serial manner, are suitable for the fabrication of few devices where cross contamination can be an issue. These tools can also be used in combination with other printing techniques, e.g. microcontact printing (to ink the stamp with different molecules), thus allowing different species to be printed in a single run in a parallel manner [GER 05].

3.5. Bibliography

[AYE 07] AYELA C., VANDEVELDE F., LAGRANGE D., *et al.*, "Combining resonant piezoelectric micromembranes with molecularly imprinted polymers", *Angewandte Chemie-International Edition,* vol. 46, pp. 9271–9274, 2007.

[BAL 05] BALAKIREV M.Y., PORTE S., VERNAZ-GRIS M.,. *et al.*, "Photochemical patterning of biological molecules inside a glass capillary", *Analytical Chemistry,* vol. 77, pp. 5474–5479, September 2005.

[BAR 98] BARIE N., RAPP M., SIGRIST H., *et al.*, "Covalent photolinker-mediated immobilization of an intermediate dextran layer to polymer-coated surfaces for biosensing applications", *Biosensors & Bioelectronics,* vol. 13, pp. 855–860, October 1998.

[BAR 05] BARRON J.A., YOUNG H.D., DLOTT D.D., *et al.*, "Printing of protein microarrays via a capillary-free fluid jetting mechanism", *Proteomics,* vol. 5, pp. 4138–4144, November 2005.

[BAR 10] BARTON K., MISHRA S., SHORTER K. A., *et al.*, "A desktop electrohydrodynamic jet printing system", *Mechatronics,* vol. 20, pp. 611–616, August 2010.

[BEL 03] BELAUBRE P., GUIRARDEL M., GARCIA G., *et al.*, "Fabrication of biological microarrays using microcantilevers", *Applied Physics Letters,* vol. 82, pp. 3122–3124, May 2003.

[BER 08] BERTHET-DUROURE N., LEÏCHLÉ T., POURCIEL J. B., *et al.*, "Interaction of biomolecules sequentially deposited at the same location using a microcantilever-based spotter", *Biomedical Microdevices,* vol. 10, pp. 479–487, August 2008.

[BER 00] BERNARD A., RENAULT J.P., MICHEL B., *et al.*, "Microcontact printing of proteins", *Advanced Materials,* vol. 12, pp. 1067–1070, July 2000.

[BIE 04a] BIETSCH A., ZHANG J.Y., HEGNER M., *et al.*, "Rapid functionalization of cantilever array sensors by inkjet printing", *Nanotechnology,* vol. 15, pp. 873–880, August 2004.

[BIE 04b] BIETSCH A., HEGNER M., LANG H.P., *et al.*, "Inkjet deposition of alkanethiolate monolayers and DNA oligonucleotides on gold: evaluation of spot uniformity by wet etching", *Langmuir,* vol. 20, pp. 5119–5122, June 2004.

[CAL 01] CALVERT P., "Inkjet printing for materials and devices", *Chemistry of Materials,* vol. 13, pp. 3299–3305, October 2001.

[CHU 05] CHUNG S.W., GINGER D.S., MORALES M.W., *et al.*, "Top-down meets bottom-up: dip-pen nanolithography and DNA-directed assembly of nanoscale electrical circuits", *Small,* vol. 1, pp. 64–69, January 2005.

[CLA 93] CLARK P., BRITLAND S., CONNOLLY P., "Growth cone guidance and neuron morphology on micropatterned laminin surfaces", *Journal of Cell Science,* vol. 105, pp. 203–212, May 1993.

[DEL 04] DELADI S., TAS N.R., BERENSCHOT J.W., *et al.*, "Micromachined fountain pen for atomic force microscope-based nanopatterning", *Applied Physics Letters,* vol. 85, pp. 5361–5363, November 2004.

[DEM 01] DEMERS L.M., PARK S.J., TATON T.A., *et al.*, "Orthogonal assembly of nanoparticle building blocks on dip-pen nanolithographically generated templates of DNA", *Angewandte Chemie-International Edition,* vol. 40, pp. 3071–3073, 2001.

[DES 07] DESCAMPS E., LEÏCHLÉ T., CORSO B., *et al.*, "Fabrication of oligonucleotide chips by using parallel cantilever-based electrochemical deposition in picoliter volumes", *Advanced Materials,* vol. 19, p. 1816, July 2007.

[DID 12] DIDAR T.F., FOUDEH A.M., TABRIZIAN M., "Patterning multiplex protein microarrays in a single microfluidic channel", *Analytical Chemistry,* vol. 84, pp. 1012–1018, January 2012.

[GER 05] GERDING J.D., WILLARD D.M., VAN ORDEN A., "Single-feature inking and stamping: a versatile approach to molecular patterning", *Journal of the American Chemical Society,* vol. 127, pp. 1106–1107, February 2005.

[GIN 04] GINGER D.S., ZHANG H., MIRKIN C.A., "The evolution of dip-pen nanolithography", *Angewandte Chemie-International Edition,* vol. 43, pp. 30–45, 2004.

[GIZ 02] GIZELI E., LOWE C.R., *Biomolecular Sensors*, Taylor & Francis, 2002.

[GOL 00] GOLDMANN T., GONZALEZ J.S., "DNA-printing: utilization of a standard inkjet printer for the transfer of nucleic acids to solid supports", *Journal of Biochemical and Biophysical Methods,* vol. 42, pp. 105–110, March 2000.

[GUE 05] GUENAT O.T., GENERELLI S., DADRAS M., *et al.*, "Generic technological platform for microfabricating silicon nitride micro- and nanopipette arrays", *Journal of Micromechanics and Microengineering,* vol. 15, pp. 2372–2378, December 2005.

[HON 99] HONG S.H., ZHU J., MIRKIN C.A., "Multiple ink nanolithography: toward a multiple-pen nano-plotter", *Science,* vol. 286, pp. 523–525, October 1999.

[HUG 01] HUGHES T.R., MAO M., JONES A. R., *et al.*, "Expression profiling using microarrays fabricated by an inkjet oligonucleotide synthesizer", *Nature Biotechnology,* vol. 19, pp. 342–347, April 2001.

[IVA 01] IVANISEVIC A., MIRKIN C.A., "'Dip-pen' nanolithography on semiconductor surfaces", *Journal of the American Chemical Society,* vol. 123, pp. 7887–7889, August 2001.

[JAS 95] JASCHKE M., BUTT H.J., "Deposition of organic material by the tip of a scanning force microscope", *Langmuir,* vol. 11, pp. 1061–1064, April 1995.

[JUN 01] JUNCKER D., SCHMID H., BERNARD A., *et al.*, "Soft and rigid two–level microfluidic networks for patterning surfaces", *Journal of Micromechanics and Microengineering,* vol. 11, pp. 532–541, September 2001.

[KIM 05] KIM K.H., MOLDOVAN N., ESPINOSA H.D., "A nanofountain probe with sub-100 nm molecular writing resolution", *Small,* vol. 1, pp. 632–635, May 2005.

[KRA 05] KRAUS T., STUTZ R., BALMER T.E., *et al.*, "Printing chemical gradients", *Langmuir,* vol. 21, pp. 7796–7804, August 2005.

[LAN 04] LANGE S.A., BENES V., KERN D.P., *et al.*, "Microcontact printing of DNA molecules", *Analytical Chemistry,* vol. 76, pp. 1641–1647, March 2004.

[LEE 03] LEE K.B., LIM J.H., MIRKIN C.A., "Protein nanostructures formed via direct-write dip-pen nanolithography", *Journal of the American Chemical Society,* vol. 125, pp. 5588–5589, May 2003.

[LEE 04] LEE K.B., KIM E.Y., MIRKIN C.A., *et al.*, "The use of nanoarrays for highly sensitive and selective detection of human immunodeficiency virus type 1 in plasma", *Nano Letters,* vol. 4, pp. 1869–1872, October 2004.

[LEI 08] LEÏCHLÉ T., LISHCHYNSKA M., MATHIEU F., *et al.*, "A microcantilever-based picoliter droplet dispenser with integrated force sensors and electroassisted deposition means", *Journal of Microelectromechanical Systems,* vol. 17, pp. 1239–1253, October 2008.

[LEI 12] LEÏCHLÉ T., LIN Y.L., CHIANG P.C., *et al.*, "Biosensor-compatible encapsulation for pre-functionalized nanofluidic channels using asymmetric plasma treatment", *Sensors and Actuators B-Chemical,* vol. 161, pp. 805–810, January 2012.

[LEW 99] LEWIS A., KHEIFETZ Y., SHAMBRODT E., *et al.*, "Fountain pen nanochemistry: atomic force control of chrome etching", *Applied Physics Letters,* vol. 75, pp. 2689–2691, October 1999.

[LIM 03] LIM J.H., GINGER D.S., LEE K.B., *et al.*, "Direct-write dip-pen nanolithography of proteins on modified silicon oxide surfaces", *Angewandte Chemie-International Edition,* vol. 42, pp. 2309–2312, 2003.

[MEI 03] MEISTER A., JENEY S., LILEY M., *et al.*, "Nanoscale dispensing of liquids through cantilevered probes", *Microelectronic Engineering,* vol. 67–68, pp. 644–650, June 2003.

[MEI 04] MEISTER A., LILEY M., BRUGGER J., *et al.*, "Nanodispenser for attoliter volume deposition using atomic force microscopy probes modified by focused-ion-beam milling", *Applied Physics Letters,* vol. 85, pp. 6260–6262, December 2004.

[MRK 95] MRKSICH M., WHITESIDES G.M., "Patterning self-assembled monolayers using microcontact printing – a new technology for biosensors", *Trends in Biotechnology,* vol. 13, pp. 228–235, June 1995.

[PIN 99] PINER R.D., ZHU J., XU F., *et al.*, "'Dip-pen' nanolithography", *Science,* vol. 283, pp. 661–663, January 1999.

[REI 05] REIS N., AINSLEY C., DERBY B., "Inkjet delivery of particle suspensions by piezoclectric droplet ejectors", *Journal of Applied Physics,* vol. 97, p. 6, May 2005.

[ROD 05] RODOLFA K.T., BRUCKBAUER A., ZHOU D.J., *et al.*, "Two-component graded deposition of biomolecules with a double-barreled nanopipette", *Angewandte Chemie-International Edition,* vol. 44, pp. 6854–6859, 2005.

[ROU 09] ROUPIOZ Y., BERTHET-DUROURE N., LEÏCHLÉ T., *et al.*, "Individual blood-cell capture and 2D organization on microarrays", *Small,* vol. 5, pp. 1493–1497, July 2009.

[SAL 05] SALAITA K., LEE S.W., WANG X.F., *et al.*, "Sub-100 nm, centimeter-scale, parallel dip-pen nanolithography", *Small,* vol. 1, pp. 940–945, September 2005.

[SAN 98] SANFORD M.S., CHARLES P.T., COMMISSO S.M., *et al.*, "Photoactivatable cross-linked polyacrylamide for the site-selective immobilization of antigens and antibodies", *Chemistry of Materials,* vol. 10, pp. 1510–1520, June 1998.

[SCH 98] SCHENA M., HELLER R.A., THERIAULT T.P., *et al.*, "Microarrays: biotechnology's discovery platform for functional genomics", *Trends in Biotechnology,* vol. 16, pp. 301–306, July 1998.

[SHE 06] SHEN S.C., PAN C.T., WANG Y.R., *et al.*, "Fabrication of integrated nozzle plates for inkjet print head using microinjection process", *Sensors and Actuators a-Physical,* vol. 127, pp. 241–247, March 2006.

[SIN 94] SINGHVI R., KUMAR A., LOPEZ G.P., *et al.*, "Engineering cell-shape and function", *Science,* vol. 264, pp. 696–698, April 1994.

[SU 05] SU M., PAN Z.X., DRAVID V.P., *et al.*, "Locally enhanced relative humidity for scanning probe nanolithography", *Langmuir,* vol. 21, pp. 10902–10906, November 2005.

[TAH 03] TAHA H., MARKS R.S., GHEBER L. A., *et al.*, "Protein printing with an atomic force sensing nanofountainpen", *Applied Physics Letters,* vol. 83, pp. 1041–1043, August 2003.

[TSA 03] TSAI J.G.F., CHEN Z.G., NELSON S., *et al.*, "A silicon-micromachined pin for contact droplet printing", *16th IEEE Annual International Conference on Micro Electro Mechanical Systems*, Kyoto, Japan, pp. 295–298, 2003.

[WAN 03] WANG X.F., RYU K.S., BULLEN D.A., *et al.*, "Scanning probe contact printing", *Langmuir,* vol. 19, pp. 8951–8955, October 2003.

[WAN 05] WANG X. F., LIU C., "Multifunctional probe array for nano patterning and imaging", *Nano Letters,* vol. 5, pp. 1867–1872, October 2005.

[XU 04] XU J.T., LYNCH M., HUFF J.L., *et al.*, "Microfabricated quill-type surface patterning tools for the creation of biological micro/nano arrays", *Biomedical Microdevices,* vol. 6, pp. 117–123, June 2004.

[XU 05] XU T., JIN J., GREGORY C., *et al.*, "Inkjet printing of viable mammalian cells", *Biomaterials,* vol. 26, pp. 93–99, January 2005.

[YAM 84] YAMASHITA M., FENN J.B., "Electrospray ion-source – another variation on the free-jet theme", *Journal of Physical Chemistry,* vol. 88, pp. 4451–4459, 1984.

[YOG 01] YOGI O., KAWAKAMI T., YAMAUCHI M., *et al.*, "On-demand droplet spotter for preparing pico- to femtoliter droplets on surfaces", *Analytical Chemistry,* vol. 73, pp. 1896–1902, April 2001.

[ZHA 02] ZHANG H., LI Z., MIRKIN C.A., "Dip-pen nanolithography-based methodology for preparing arrays of nanostructures functionalized with oligonucleotides", *Advanced Materials,* vol. 14, pp. 1472, October 2002.

[ZHA 03] ZHANG H., LEE K. B., LI Z., *et al.*, "Biofunctionalized nanoarrays of inorganic structures prepared by dip-pen nanolithography", *Nanotechnology,* vol. 14, pp. 1113–1117, October 2003.

[ZHA 04] ZHANG H., ELGHANIAN R., AMRO N.A., *et al.*, "Dip pen nanolithography stamp tip", *Nano Letters,* vol. 4, pp. 1649–1655, September 2004.

4

From MEMS to NEMS Biosensors

Micro electromechanical systems-based biosensors encompass all biosensors involving the mechanical transduction of biological recognition events happening on a miniaturized free-standing structure. With the recent advances of nanotechnologies, a new class of nanoscale resonators has emerged with improved performances in terms of sensitivity and minimum detectable mass. However, to date, these devices have not been successfully used for real-time biosensing. This chapter aims to present and discuss the challenges preventing the breakthrough of NEMS biosensors and provides some tentative answers.

4.1. Importance of downscaling

The apparition and spreading of nanopatterning and nanofabrication techniques in research laboratories and institutes has quickly opened the door to the realization of free-standing nanoscale structures promising to fulfill various application needs. A little more than 10 years ago, three major papers simultaneously laid down the foundations of the "nanoelectromechanical systems for biology" (or bioNEMS) realm: while Roukes [ROU 01] and Craighead [CRA 00] separately paved the way for brand-new nanoelectromechanical systems in the United States, a European joint effort (IBM Zurich – University of Basel) [FRI 00] demonstrated the potential impact of translating biomolecular recognition into nanomechanics.

The main importance of downscaling MEMS structures operated in dynamic mode for sensing applications is a significant increase in the sensor's ultimate performances (sensitivity and minimum detectable mass (MDM)). Indeed, the sensitivity of a resonant NEMS biosensor is directly proportional to its resonant frequency and inversely proportional to its mass. This immediately implies that a highly sensitive mechanical biosensor is a structure of nanoscale dimensions (i.e. of very low weight – in the order of femtogram – and very high natural resonant frequency – in the order of megahertz). In addition, providing that there is no energy dissipation while vibrating, such a structure can oscillate with a high-quality factor (on the order of 1,000–10,000), thus enabling determination of the resonant frequency with a high accuracy and so measurement of a mass with a high sensitivity. In addition, by taking advantage of silicon-based technology for fabricating new generations of extremely miniaturized mechanical sensors, major needs of the *in vitro* diagnostic industry such as label-free, multiplexed, real-time, reliable biosensing operations can be successfully fulfilled by NEMS biosensors.

Yet, to this day, there has not been a single example of an NEMS-based truly operational biosensor: indeed, all hopes of using these amazing structures literally vanished when confronted with the reality on the ground. It quickly appeared that when combining the biological requirements with the technological constraints at the nanoscale, the complexity level increased exponentially and brought more issues than solutions even when dealing with basic problems. Runners in the bioNEMS competition had to solve the "N nanosensors detecting N different species in a liquid medium" equation. This involves simultaneously dealing with the analysis of heterogeneous analytes, the associated difficulties to discriminate specific from non-specific interactions as well as operating ultradense arrays of nanodevices that require individual addressing in terms of actuation/sensing/biofunctionalization.

Eventually, rather than treating all items as a block, proofs of concept that represent only part of the required solution have been proposed so far (e.g. one nanosensor detecting ONE kind of biological species in vacuum [ILI 05]).

To fully take advantage of the potential of NEMS for biosensing applications, there are thus amazing challenges to overcome. Here, we list the most important issues that are faced by the development of NEMS biosensors.

4.2. Challenges faced by NEMS for biosensing applications

For resonant NEMS to be considered as a viable alternative to their actual biosensing macro counterparts, they have to simultaneously meet three major requirements: high mass responsivity (MR), low MDM and low response time (RT).

Without any doubt, as emphasized by theoretical studies [EKI 04], the two first specifications (high MR and low MDM) can theoretically and successfully be addressed by NEMS devices if they exhibit ultra-high quality factors. Such predictions have already been validated in cases of virus sensing [ILI 04], enumeration of DNA molecules [ILI 05], single molecule nanomechanical mass spectrometry [NAI 09] and even chemical sensing in gaseous environment (Figure 4.1) [BAR 12].

However, in order to target real-time analysis in crude samples as, for instance, in the case of point-of-care analysis of physiological liquids, measurements must be carried out in liquid environment: this leads to a drastic decrease in the quality factors of the nanostructures presented so far, i.e. nanocantilevers, preventing them from being applicable. Moreover, NEMS must be adequately functionalized to function as biosensors, and so far there is a severe lack of reliable and simple technical solutions for the functionalization of nanoscale free-standing structures with multiplexing capability. Finally, and hopelessly, nanometer scale sensors have also been proved, still in theory [NAI 06], to be inadequate to practical RT scales because of diffusion limitations or inefficient mass transport, which, if confirmed as such, could definitely impede the way toward realistic biosensing applications: it is here that packaging comes to play a prominent role.

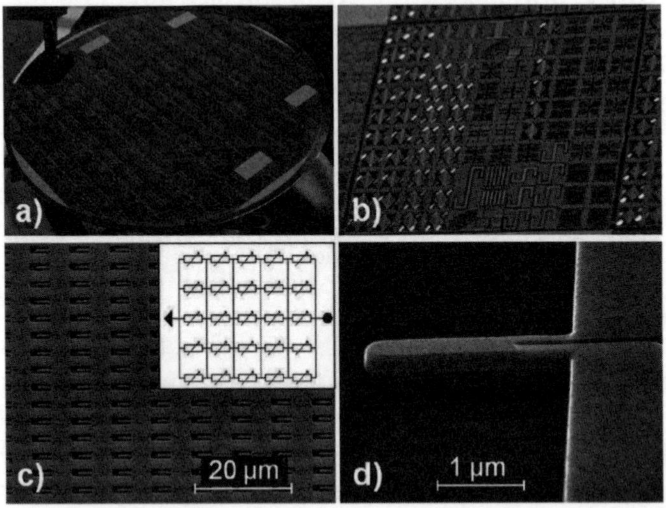

Figure 4.1. *a) Photograph of a full 200 mm wafer with patterned NEMS arrays. b) Zoomed-in photograph of one 20 mm wafer die containing a variety of nanofabricated resonator array structures. c) Scanning electron micrograph of a section of a cantilever array. Inset: Schematic of a combined series–parallel electrical connection of array elements. d) Scanning electron micrograph (oblique view) of an individual array component [BAR 12]*

In the following sections, we will attempt to provide preliminary answers to these issues (without mentioning other critical aspects required for multiplexed operation of large arrays of NEMS, e.g. those related to the associated read-out electronics) that could lead, in the near future, to the emergence of a generation of NEMS biosensors ready to tackle practical applications.

4.2.1. *Issues related to nanomechanical transducers*

The common ground of the great majority of publications in the field of NEMS is the use of beam-based mechanical structures, either in a cantilever or bridge-like configuration. Nevertheless, when it comes to using beams as resonant biosensors, they appear limited by relative brittleness with respect to the functionalization step and, most importantly, they exhibit low quality factors (less than 50) in liquid media [VAZ 09]. This aspect, related to the damping of the beams oscillation in liquid, directly impacts the MDM and subsequently, the signal-to-noise ratio of the sensor.

One effective way to tackle those geometry-related issues relies on considering alternative designs and/or shifts in operating configurations. While the latter solution has already been implemented by using lateral mode resonant cantilevers that result in Q factors up to 67 in water [BEA 10], it seems that the most promising way is to explore the potential of nanomembrane-like geometries. Indeed, in the past, extensive work has proved that membrane geometries allow reaching high-quality factors in liquid due to a dynamic behavior unchanged by the liquid viscosity within certain limits (Figure 4.2) [AYE 07, ALA 10]. It is important to note here that this type of geometry is, moreover, better adapted to the integration of additional layers used for mechanical actuation, displacement sensing and biorecognition purposes. Although not discussed here, the integration of efficient actuation/sensing schemes and materials is not as straightforward for nanosized mechanical structures as it is for their macroscale counterparts and is now subject to intense research works.

The miniaturization of micromembranes down to the nanoscale requires the use of nanofabrication tools and a shift in design methodology: the fabrication of micromembranes that classically relies on deep reaction ion etching (RIE) of the backside of a silicon-on-insulator substrate [NIC 05] is not suitable for downscaling. Thus, suspended nanostructures can be obtained using a sacrificial layer [WIL 11]. However, care must be taken during the release process in order to avoid collapsing of the structure due to capillary forces, and the main drawback of this approach is the difficulty to fabricate fully sealed membranes, in order to avoid issues when operating these devices in liquid with integrated sensing and actuation schemes. Hence, there is an urgent need in proposing new fabrication approaches for the realization of NEMS membranes. Recently, the use of an interesting alternative technology, the so-called transfer-printing [KIM 10], was proposed for the simple fabrication in a few steps of free-standing nanomembranes by means of micromasonry [KEU 13]. However, these devices still need to be proven adequate for proper oscillation within liquids.

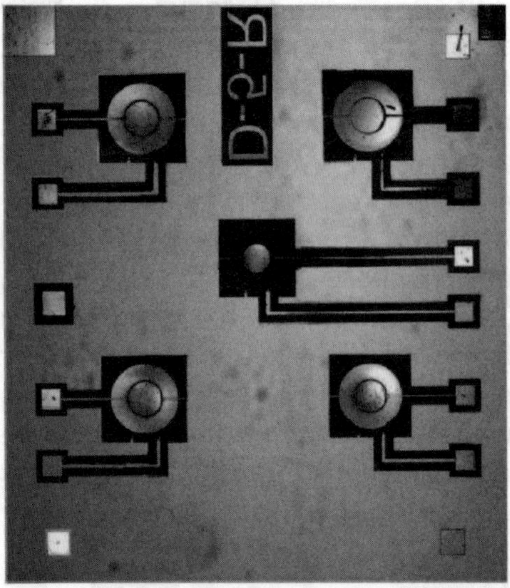

Figure 4.2. *Optical photograph of a fabricated chip containing an array of five resonating 440 μm diameter micromembranes with integrated actuation and sensing schemes by means of a piezoelectric thin film [NIC 12]*

4.2.2. *Issues related to the functionalization of NEMS*

One of the major issues preventing the breakthrough of NEMS devices for biodetection applications remains the biofunctionalization of the sensor, i.e. the grafting of probe molecules onto its active surface [ARL 11]. While this technical difficulty can be easily circumvented when taking a bottom-up approach since the nanostructures can be functionalized before integrating them into a device, it raises tremendous constraints when considering NEMS devices obtained by means of micro- and nanofabrication techniques. Indeed, preservation of the bioactivity of the sensing layer imposes the avoidance of subsequent fabrication steps leading to the degradation or the alteration of the probe molecules, e.g. processes involving plasma and vacuum environments or harsh chemicals [LEÏ 12]. This means that surface modification is most likely to be carried out at the end of the fabrication process, i.e. after the release of the nanostructures. Hence, care must be taken when delivering probe molecules onto free-standing fragile structures to keep the integrity of the sensor.

Moreover, the functionalization of a large array of NEMS devices implies that submicron patterns can be created on a at large scale. Thus, the most common functionalization method relying on the immersion of the entire sample into a dedicated biochemical solution does not fulfill the need to create localized biofunctionalized areas [FIS 08]. Many of the techniques that have been presented in Chapter 3 to locally biofunctionalize microscale sensors, such as inkjet printing [BIE 04] and the use of fluid patterning by means of separate microcapillaries [MCK 02], are either not likely to produce sub-micron patterns or to be scaled up to create large and multiplexed arrays. Also, the photolithographic light-directed synthesis method [PEA 94] that is already used for gene chip fabrication is indeed appropriate for large multiplexed arrays, but the diffraction limit makes it difficult to apply to nanoscale patterning. Liquid drop dispensers that have been demonstrated as useful tools for the biofunctionalization of MEMS sensors [VAN 07] are not suitable because the capillary forces induced when retracting the tool from the surface are most likely large enough to damage nanostructures. Other technologies used to create nanoscale patterns by means of tip-based liquid or molecular transfer, such as miniaturized fountain pen [SAY 07] and scanning probes, could possibly address the issues related to the biofunctionalization of fragile nanostructures, even if the creation of patterns on large areas seems technically difficult to implement. Finally, we can think of using interesting approaches relying on the use of localized electrochemical reactions induced at the surface of addressable electrodes [HAR 07, DES 07], despite the foreseen problems to be raised by the high-multiplexing requirements of the electrode array in order to access individual or groups of nanostructures.

Given the overall context of the localized deposition of biomolecules on solid surfaces, it has recently been proposed to use microcontact printing (µCP) [KUM 93], which is a highly parallel deposition technology [DEC 04] with nanoscale resolution [GU 08], for the localized biofunctionalization of large-scale arrays of free-standing nanocantilevers. µCP is a "dry" deposition technology known to keep biomolecules fully active after deposition. The proposed method relied on a modified µCP process, where the cavities in a polydimethylsiloxane (PDMS) stamp are actually used to transfer the molecules of interest, while the protrusion of the stamp provides mechanical support during the printing step and can additionally be used to deposit blocking molecules on the non-reactive parts of the array [GUI 12]. The stamp is designed so that when placed onto a surface, the bottom of the cavities comes into contact with that surface if a pressure is applied, thus allowing molecular transfer without inducing too much force onto the structure to functionalize. The functionalization process, illustrated in Figure 4.3, includes two stages: the PDMS stamp preparation and inking and the actual printing or transfer process.

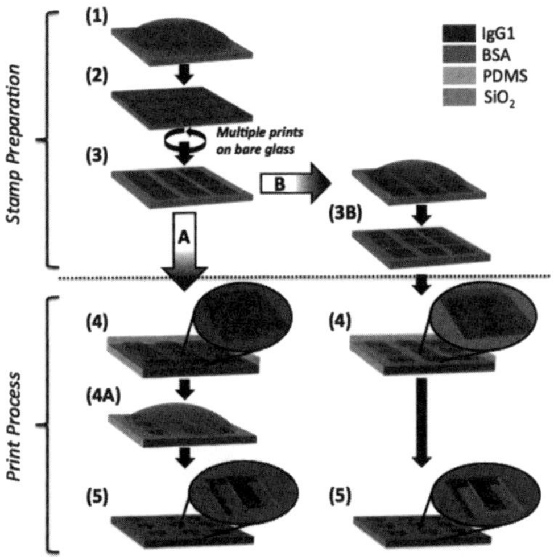

Figure 4.3. *Schematics of the NEMS functionalization process using a PDMS stamp; two approaches can be used: (A) Surface functionalization by microcontact printing followed by blocking the remaining surfaces with a batch process. (B) Surface functionalization and passivation both carried out by microcontact printing. (1) Inking the stamp with the desired molecules (e.g. IgG1). (2) Washing and drying the stamp. (3) Cleaning outside the stamp grooves, i.e. removing the molecules from the stamp base, by carrying multiple prints. (3B) Inking the stamp with the antifouling molecules (BSA). (4) Printing after aligning the stamp and the chip. (4A) Incubating the chip with the antifouling molecule (BSA). (5) Result of the NEMS biofunctionalization process [SAL 12]*

The functionalization of arrays of free-standing nanocantilevers as dense as 10^5 nanostructures/cm^2 was recently demonstrated using the modified μCP technique (Figure 4.4). The proper bioactivity of an antibody deposited onto the nanocantilevers and the effective blocking property of a bovine serum albumin (BSA) layer were both assessed by incubating specific and non-specific tagged secondary antibodies followed by fluorescence imaging. Furthermore, measurement of the resonant frequency of the nanocantilevers before and after functionalization and biological recognition demonstrated that using μCP for device functionalization did not damage the nanostructures and preserved the mechanical sensing capability of the

NEMS, thus providing a solid technical solution for the functionalization of NEMS biosensors [SAL 12].

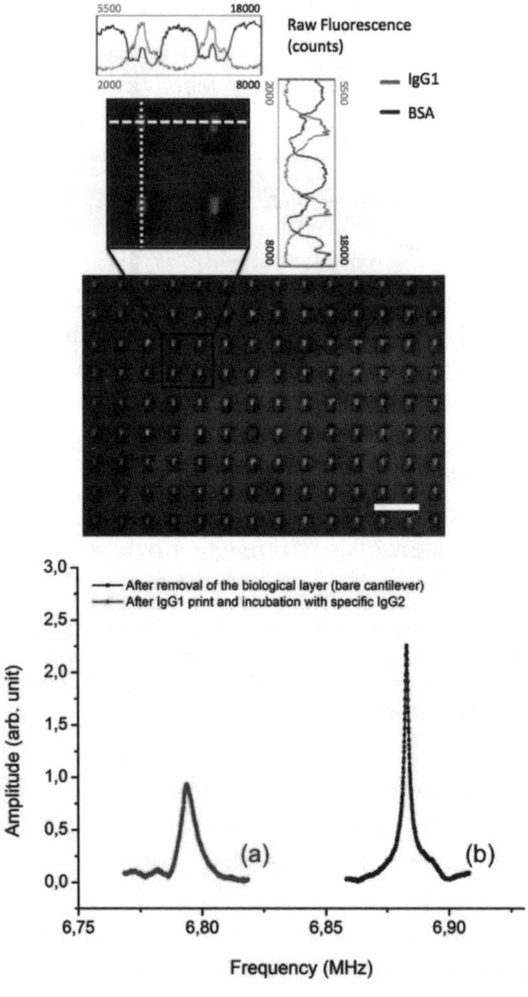

Figure 4.4. *Top: An array of nanocantilevers biofunctionalized using the μCP patterning technique [SAL 12]. Superimposed fluorescence pictures of an array of nanocantilevers functionalized after printing an Alexa Fluor 660 IgG and Fluorescein-BSA in a single step (scale bar: 50μm). The inset provides a zoom of the array along with the fluorescence profiles. Bottom: Resonant frequency curves of a single nanocantilever (a) after IgG1 print followed by interaction with a specific IgG2 and (b) after removing all the printed and captured biomolecules [SAL 12]*

4.2.3. *On the importance of packaging and sample preparation*

Another concern inherent to sensor miniaturization is reaching the ultimate detection limit, i.e. detecting a single or a few copies of target, while preserving an acceptable RT. The time it takes for a biological target to find the active surface of the biosensor and cause a measurable event depends upon scattering and diffusion effects under poor mass transport conditions. Thus, it is clear that the smaller the capture surface is, the longer it takes for a target to interact with the sensor, and in other words, the longer the sensor RT is. In 2005, Sheehan and Whitman theoretically demonstrated that in the diffusion-limited regime, in a 1 fM solution, it takes several hours for a 20 nucleotides double-stranded DNA (with a diffusion constant of 150 µm²/s) to reach a nanoscale sensor whereas this time is reduced to few tens of seconds in the case of a 100 µm diameter hemispherical sensor (Figure 4.5) [SHE 05]. This demonstrates the antagonism between the potential of NEMS biosensors for ultrasensitive detection and dynamic response incompatible with practical assay timescales.

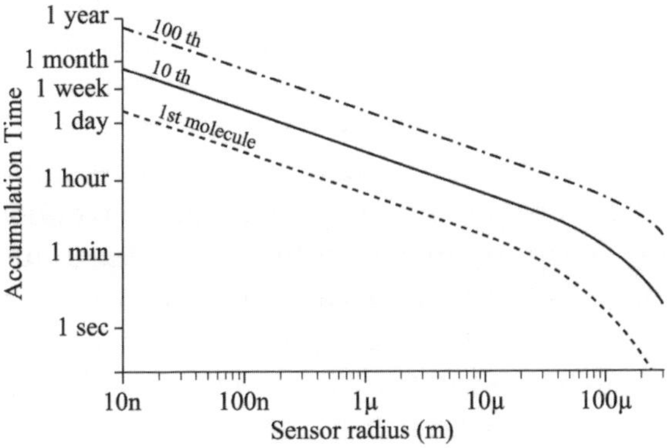

Figure 4.5. *Time required for a hemisphere sensor to accumulate 1, 10 and 100 molecules (with a diffusion constant of 150 µm²/s, characteristic of single-stranded DNA approximately 20 bases long) via static diffusion when submerged in a semi-infinite 1 fM solution. For radii smaller than 10 µm, the required time varies linearly with the radius [SHE 05]*

One way to face this trade-off is to enhance mass transport of target molecules. This can be accomplished by increasing the volumetric flow rate via pressure or electrokinetic-driven flow using dedicated fluidic setups. However, increasing the flow rate does not necessarily greatly enhance target flux to the sensor surface, since this enhancement is greater on larger area sensors than on nanoscale devices. This is because there is a degree of transient enhancement for larger area sensors at earlier time points, due to the high concentration gradients that deplete the solution, whereas within nanoscale sensors the degree of target depletion in the solution is minimal.

Thus, the easiest route toward a practical assay time at ultra-low sample concentration is to carry out on-chip target preconcentration before performing the analysis in order to increase the concentration of the analyte to detect in the vicinity of the sensor. For this aim, several techniques have lately been developed in micro- and nanofluidic devices. For instance, electrokinetic methods, which are based on the manipulation of electrical double layer around the biomarker, have been demonstrated to achieve localized biomarker enrichment at million-fold levels [WAN 05], especially due to the large volume reduction and the utilization of ion concentration polarization at nanofluidic device interfaces [LEE 08]. More recently, the concept of a nanofluidic molecular dam was presented where electrokinetic forces and negative dielectrophoresis are, respectively, used to drive molecules toward and repulse them out of a dielectric nanoconstriction, resulting in a fast accumulation of targets close to the constriction. Enrichment of proteins, such as streptavidin, to million-fold levels within a few seconds was successfully demonstrated by using this technique [LIA 12].

Another way to counteract the negative effect of active area miniaturization on the sensor RT is to choose a trade-off strategy that consists of gaining advantage by considering a single NEMS device not as being alone but as part of a functional array of similar devices [SAM 11]. This paradigm allows us, while preserving the benefits of high MR and low MDM of a single device, to use the considerably higher capture area of the NEMS array such that the RT reaches practical relevance. However, in that case, the non-reactive areas of the chip containing multiple sensors functionalized with a single type of probe molecule must be adequately coated with an antifouling film in order to lower the probability of adsorption of the few copies of target molecules anywhere else than on sensitive areas and hence to permit their detection at ultra-low concentration levels.

Still, from a practical point of view, a related problem arises from the need to detect targets at such low concentration levels (<ng/ml) within physiological media that contain several other molecules (proteins) at orders of magnitude which are of higher concentration levels (>mg/ml) [AND 02]. Hence, while miniaturized biosensor platforms can detect these biomarkers at the required sensitivity levels in the absence of interfering proteins, they are not relevant within a clinical setting if they exhibit even a ppm–ppb-level response toward the background proteins. Furthermore, this highlights the need for methods to selectively enrich the concentration of biomarkers over other proteins in biological fluids, in proximity of the sensor immobilized with capture probes.

Nevertheless, the prerequisite to afford a truly portable biosensing device for, e.g., point-of-care applications requires the sample preparation to be carried out on-chip prior to analysis. This necessity to integrate sample preparation to the sensing platform, which is common to all types of integrated biosensors, has been legitimately pointed out as being the “weak link in microfluidic-based biodetection” [MAR 08]. Tentative answers on how to replace and integrate the various steps used to prepare a biological sample in a medical laboratory, e.g. centrifugation, purification or extraction, on a chip are to be found in the fields of lab-on-a-chip [MAR 10] and micro total analysis systems (µTAS) [REY 02, AUR 02]. There are now dedicated technological solutions that are being implemented to reproduce each step used in any sample preparation process flow on a chip: merging these technical bricks with miniaturized MEMS and NEMS sensors is on its way.

4.3. Economic considerations

In the context of *in vitro* diagnostics (IVDs), nanotechnology is used for the development of novel classes of biosensors aiming to improve and make existing biosensing schemes more sensitive, to pave the way toward point-of-care applications or to develop completely new diagnostic test platforms. As a whole, nanotechnology-enabled IVDs can be broadly divided into two main approaches: (1) use of nanoparticles as markers for biomolecules and (2) novel sensor platforms using free-standing nanostructures for biological applications.

Commercialization efforts in nanotechnology-enabled IVDs are picking up worldwide. A recent study identified more than 30 businesses (mostly small and medium enterprises (SMs) and start-ups) having nanotechnology-enabled IDV products on the market, which is remarkable when considering that the first publications in this field appeared only in the late 1990s [WAG 08]. The corresponding sales volume was estimated at more than € 600 million at the end of 2010, which represents only 2.7% of the whole IVDs potential market environment.

When focusing on biosensing platforms using free-standing micro- and nanostructures, it appears that the field is still in a fairly early development stage and that only applications in biomedical research are currently found. Yet, there are a couple of companies, such as Bioforce Nanosciences (USA) or Concentris (Switzerland), that further develop these systems for clinical applications. BioNEMS start-ups will face significant roadblocks including the regulatory process for medical products, the skepticism of investors and poorly focused business plans, the latter point already having been the origin of the bankruptcy of start-ups such as Cantion (Denmark), Sensia (Spain) or Protiveris (USA). Among those, the hugest challenge for bioNEMS to succeed in the nanotechnology-enabled IVDs market is the pharmaceutical industry's reluctance to invest significant amounts of money into such systems as one of the next big endeavors. Yet, being committed to delivering higher earnings year after year, pharmaceutical companies have to deliver health benefits while producing at lower costs or they have to make R&D much more productive.

Bringing bioNEMS-related technology complexity to lowest possible levels will allow addressing the previous issues while improving the speed through the regulatory processes and lowering the time-to-market value.

4.4. Bibliography

[ALA 10] ALAVA T., MATHIEU F., MAZENQ L., *et al.*, "Piezoelectric–actuated, piezoresistive-sensed circular micromembranes for label-free biosensing applications", *Applied Physics Letters*, vol. 97, pp. 093703, 2010.

[AND 02] ANDERSON N.L., ANDERSON N.G., "The human plasma proteome – history, character, and diagnostic prospects", *Molecules & Cellular Proteomics*, vol. 1, pp. 845–867, 2002.

[ARL 11] ARLETT J.L., MYERS E.B., ROUKES M.L., "Comparative advantages of mechanical biosensors", *Nature Nanotechnology*, vol. 6, pp. 203–215, 2011.

[AUR 02] AUROUX P.-A., IOSSIFIDIS D., REYES D.R., *et al.*, "Micro total analysis systems. 2. Analytical standard operations and applications", *Analytical Chemistry*, vol. 74, pp. 2637–2652, 2002.

[AYE 07] AYELA C., NICU L., "Micromachined piezoelectric membranes with high nominal quality factors in Newtonian liquid media: a Lamb's model validation at the microscale", *Sensors and Actuators Chemical*, vol. B123, pp. 860–868, 2007.

[BAR 12] BARGATIN I., MYERS E.B., ALDRIDGE J.S., *et al.*, "Large-Scale Integration of nanoelectromechanical systems for gas sensing applications", *Nano Letters*, vol. 12, pp. 1269–1274, 2012.

[BEA 10] BEARDSLEE L.A., DEMIRCI K.S., LUZINOVA Y., *et al.*, "Liquid-phase chemical sensing using lateral mode resonant cantilevers", *Analytical Chemistry*, vol. 82, pp. 7542–7549, 2010.

[BIE 04] BIETSCH A., ZHANG J.Y., HEGNER M., *et al.*, "Rapid functionalization of cantilever array sensors by inkjet printing", *Nanotechnology*, vol. 15, pp. 873–880, 2004.

[CRA 00] CRAIGHEAD H.G., "Nanoelectromechanical systems", *Science*, vol. 290, pp. 1532–1535, 2000.

[DEC 04] DECRE M.M.J., SCHNEIDER R., BURDINSKI D., *et al.*, "Wave printing (I): towards large-area, multilayer microcontact printing", *Nontraditional Approaches to Patterning*, pp. 59–61, 2004.

[DES 07] DESCAMPS E., LEÏCHLÉ T., CORSO B., *et al.*, "Fabrication of oligonucleotide chips by using parallel cantilever-based electrochemical deposition in picoliter volumes", *Advanced Materials*, vol. 19, pp. 1816–1821, 2007.

[EKI 04] EKINCI K.L., YANG Y.T., ROUKES M.L., "Ultimate limits to inertial mass sensing based upon nanoelectromechanical systems", *Journal of Applied Physics*, vol. 95, pp. 2682–2689, 2004.

[FIS 08] FISCHER L.M., WRIGHT V.A., GUTHY C., *et al.*, "Specific detection of proteins using nanomechanical resonators", *Sensors and Actuators Chemical*, vol. B134, pp. 613–617, 2008.

[FRI 00] FRITZ J., BALLER M.K., LANG H.P., *et al.*, "Translating biomolecular recognition into nanomechanics", *Science*, vol. 288, pp. 316–318, 2000.

[GU 08] GU J., XIAO X., TAKULAPALLI B.R., MORRISON M.E., *et al.*, "A new approach to fabricating high density nanoarrays by nanocontact printing", *Journal of Vacuum Science and Technology*, vol. B26, pp. 1860–1865, 2008.

[GUI 12] GUILLON S., SALOMON S., SEICHEPINE F., *et al.*, "Biological functionalization of massively parallel arrays of nanocantilevers using microcontact printing", *Sensors and Actuators Chemical*, vol. B161, pp. 1135–1138, 2012.

[HAR 07] HARPER J.C., POLSKY R., DIRK S.M., *et al.*, "Electroaddressable selective functionalization of electrode arrays: catalytic NADH detection using aryl diazonium modified gold electrodes", *Electroanalysis*, vol. 19, pp. 1268–1274, 2007.

[ILI 04] ILIC B., YANG Y., CRAIGHEAD H.G., "Virus detection using nanoelectromechanical devices", *Applied Physics Letters*, vol. 85, pp. 2604–2606, 2004.

[ILI 05] ILIC B., YANG Y., AUBIN K., *et al.*, "Enumeration of DNA molecules bound to a nanomechanical oscillator", *Nano Letters*, vol. 5, pp. 925–929, 2005.

[KEU 13] KEUM H., ZHANG Y., DEZEST D., *et al.*, KIM S., "Micro-masonry for small batch processing of suspended MEMS structures", *Workshop on Enabling Nanofabrication for Rapid Innovation (ENRI 2013)*, Napa, CA, 18–21 August 2013.

[KIM 10] KIM S., WU J.A., CARLSON A., *et al.*, "Microstructured elastomeric surfaces with reversible adhesion and examples of their use in deterministic assembly by transfer printing", *Proceedings of the National Academy of Sciences of the United States of America*, vol. 107, pp. 17095–17100, 2010.

[KUM 93] KUMAR A., WHITESIDES G.M., "Features of gold having micrometer to centimeter dimensions can be formed through a combination of stamping with an elastomeric stamp and an alkanethiol ink followed by chemical etching", *Applied Physics Letters*, vol. 63, pp. 2002–2004, 1993.

[LEE 08] LEE J.H., SONG Y.-A., HAN J., "Multiplexed proteomic sample preconcentration device using surface-patterned ion-selective membrane", *Lab on a Chip,* vol. 8, pp. 596–601, 2008.

[LEÏ 12] LEÏCHLÉ T., LIN Y.-L., CHIANG P.-C., *et al.*, "Biosensor-compatible encapsulation for pre-functionalized nanofluidic channels using asymmetric plasma treatment", *Sensors and Actuators Chemical*, vol. B161, pp. 805–810, 2012.

[LIA 12] LIAO K.T., CHOU C.F., "Nanoscale molecular traps and dams for ultrafast protein enrichment in high-conductivity buffers", *Journal of the American Chemical Society*, vol. 134, pp. 8742–8745, 2012.

[MAR 08] MARIELLA R., "Sample preparation: the weak link in microfluidics-based biodetection", *Biomedical Microdevices*, vol. 10, pp. 777–784, 2008.

[MAR 10] MARK D., HAEBERLE S., ROTH G., *et al.*, "Microfluidic lab-on-a-chip platforms: requirements, characteristics and applications", *Chemical Society Reviews,* vol. 39, pp. 1153–1182, 2010.

[MCK 02] MCKENDRY R., ZHANG J.Y., ARNTZ Y., *et al.*, "Multiple label-free biodetection and quantitative DNA-binding assays on a nanomechanical cantilever array", *Proceedings of the National Academy of Sciences of the United States of America*, vol. 99, pp. 9783–9788, 2002.

[NAI 09] NAIK A.K., HANAY M.S., HIEBERT W.K., *et al.*, "Towards single-molecule nanomechanical mass spectrometry", *Nature Nanotechnology*, vol. 4, pp. 445–450, 2009.

[NAI 06] NAIR P.R., ALAM M.A., "Performance limits of nanobiosensors", *Applied Physics Letters*, vol. 88, p. 233120, 2006.

[NIC 05] NICU L., GUIRARDEL M., CHAMBOSSE F., *et al.*, "Resonating piezoelectric membranes for microelectromechanically based bioassay: detection of streptavidin-gold nanoparticles interaction with biotinylated DNA", *Sensors and Actuators Chemical*, vol. B110, pp. 125–136, 2005.

[NIC 12] NICU L., ALAVA T., LEÏCHLÉ T., *et al.*, "Integrative technology-based approach of microelectromechanical systems (MEMS) for biosensing applications", *IEEE 2012 Annual International Conference of the IEEE Engineering in Medicine and Biology Society (EMBS 2012)*, San Diego, CA, pp. 4475–4478, 28 August–1 September 2012.

[PEA 94] PEASE A.C., SOLAS D., SULLIVAN E.J., *et al.*, "Light-generated oligonucleotide array for rapid DNA sequence analysis", *Proceedings of the National Academy of Sciences of the United States of America*, vol. 91, pp. 5022–5026, 1994.

[REY 02] REYES D.R., IOSSIFIDIS D., AUROUX P.-A., *et al.*, "Micro total analysis systems. 1. Introduction, theory, and technology", *Analytical Chemistry*, vol. 74, pp. 2623–2636, 2002.

[ROU 01] ROUKES M.L., "Nanoelectromechanical systems", *11th International Conference on Solid–State Sensors and Actuators (Transducers '01)*, Munich, Germany, pp. 658–661, 10–14 June 2001.

[SAL 12] SALOMON S., LEÏCHLÉ T., DEZEST D., *et al.*, "Arrays of nanoelectromechanical biosensors functionalized by microcontact printing", *Nanotechnology*, vol. 23, p. 495501, 2012.

[SAM 11] SAMPATHKUMAR A., EKINCI K.L., MURRAY T.W., "Multiplexed optical operation of distributed nanoelectromechanical systems arrays", *Nano Letters*, vol. 11, pp. 1014–1019, 2011.

[SAY 07] SAYA D., LEÏCHLÉ T., POURCIEL J.B., *et al.*, "Collective fabrication of an in-plane silicon nanotip for parallel femtoliter droplet deposition", *Journal of Micromechanics and Microengineering*, vol. 17, pp. N1–N5, 2007.

[SHE 05] SHEEHAN P.E., WHITMAN L.J., "Detection limits for nanoscale biosensors", *Nano Letters*, vol. 5, pp. 803–807, 2005.

[VAN 07] VANDEVELDE F., LEÏCHLÉ T., AYELA C., *et al.*, "Direct patterning of molecularly imprinted microdot arrays for sensors and biochips", *Langmuir*, vol. 23, pp. 6490–6493, 2007.

[VAZ 09] VAZQUEZ J., RIVERA M.A., HERNANDO J., *et al.*, "Dynamic response of low aspect ratio piezoelectric microcantilevers actuated in different liquid environments", *Journal of Micromechanics and Microengineering*, vol. 19, p. 015020, 2009.

[WAG 08] WAGNER V., HÜSING B., GAISSER S., JRC European Commission Report EUR 23494 EN, 2008. Available at http://ftp.jrc.es/EURdoc/JRC46744.pdf.

[WAN 05] WANG Y.-C., STEVENS A.L., HAN J., "Million-fold preconcentration of proteins and peptides by nanofluidic filter", *Analytical Chemistry*, vol. 77, pp. 4293–4299, 2005.

[WIL 11] WILSON-RAE I., BARTON R.A., VERBRIDGE S.S., *et al.*, "High-Q nanomechanics via destructive interference of elastic waves", *Physical Review Letters*, vol. 106, p. 047205, 2011.

5

Comparing Performances of Biosensors: Impossible Mission?

The review of biosensing platforms provided in Chapter 2, though non-exhaustive, raises a major question: what are the comparison criteria for all those systems? In this final concluding chapter, we try to answer this crucial question in order to provide means to assess the interest and advantages of MEMS biosensors.

The classification of biosensors presented in the Introduction provides ideas about the criteria that can be used to compare the performances of the various biosensing methods. Indeed, we can take into account either technical or biological considerations.

If commercial systems had to be compared on the basis of technical considerations, we would think that this is a "lucky" situation. Indeed, data sets are provided when purchasing a commercial system; these deal generally with specifications concerning the analyte flow profile and rate ranges, typical sample volumes, working temperature, sensitivity, resolution of measurement and so forth. Yet even if these data are comprehensive and precise enough, most of the time they are difficult to compare since they are related to a specific mode of transduction, a particular species to be detected (proteins, small molecules, cells, etc.), a given functionalization strategy of the core sensor, etc.; let alone the non-commercial systems that are often insufficiently qualified.

What about biological considerations? To take into account biological criteria, the different biosensing systems should be compared in a specific detection case, for one given analyte (i.e. one kind of antibody). It seems that this second approach meets, if not all the requirements, at least the most important requirements in order to have a better idea concerning the origin of the differences in terms of performances of various systems.

A relevant example of comparing known biosensing systems is provided by Glass *et al.* [GLA 07]. The authors compare the least detectable concentration (LDC) and dynamic range (DR) of three immunoassay systems using four distinct antibodies (all specific for the same analyte but with different affinities) on each system. The evaluated systems include the industry standard (enzyme-linked immunosorbent assay (ELISA)) and two biosensing platforms (surface plasmon resonance and kinetic exclusion). The measurements of inhibition curves (response vs. analyte concentration) were contracted to outside experts or biosensor manufacturers, each of whom was supplied with the same blind samples. The biosensor manufacturers reported an estimate of the equilibrium dissociation constant (K_d) for each of the antibodies. The LDC and DR observed for the kinetic exclusion biosensor were consistent with an interpretation of K_d limited detection while that from the other biosensor and ELISA showed limits of detection above those expected for K_d limited performances. The LDC and DR of each biosensor were thus self-consistent in the sense that none of the inhibition data contradicted theoretical limits associated with the K_d as measured on that system; however, some contradictions were apparent across platforms. The use of multiple antibodies of differing K_d

improved confidence such that the observed differences in performances were associated with the immunoassay system rather than the particular analyte.

Another example encompassed by the "biological considerations" criterion is the EILATox-Oregon Biomonitoring Workshop organized in 2004 [OSH 04] that allowed gathering a diverse set of technologies for environmental toxicant detection. The aim was to challenge the notion that bioassay approaches cannot be used reliably within the safe confines of one's own laboratory. The blind samples that were provided to each of the participating groups consisted of a diverse array of toxicants at toxic concentrations. Blind agent samples in 25 ml aliquots were prepared by dissolving the chemicals in synthetic water. A list of the toxic agents was previously available to the participants, but the toxicant identities of the blind samples were unknown until the end of the workshop. Both commercially available kits as well as prototype assays were included in the workshop. Several noteworthy observations emerged from the workshop findings: even though it was problematic to directly compare the sensitivity of the biomonitoring technologies (for two reasons: some groups could only process a limited number of samples during the workshop due to time, material or travel constraints; and there was not complete uniformity with respect to the dilution levels of toxicant samples across the different technologies), no false positives were reported across any of the technologies, regardless of the stage of maturity.

A final example where technical and biological considerations are combined to show how well microelectromechanical system (MEMS) could perform where standard techniques are only partially satisfactory is shown in Figure 5.1. In this case, label-free and label-based techniques are used to measure the levels of prostate-specific antigen (PSA) in various samples (starting from spiked buffers to real human sera): PSA is a serine protease produced by the prostate (in male) or the breast (in female) epithelium and considered as positive oncological marker of prostate cancer if found above 4 ng/ml (male serum) or 1 fg/ml (female serum).

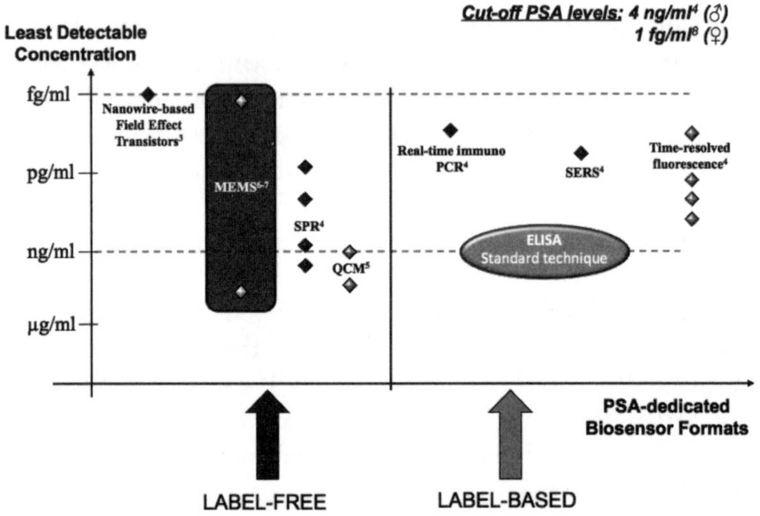

Figure 5.1. *Examples of biosensor formats classified upon the least detectable concentration applied to PSA detection (MEMS, microelectromechanical systems; SPR, surface plasmon resonance; QCM, quartz crystal microbalance; PCR, polymerase chain reaction; ELISA, enzyme-linked immunosorbent assay; SERS, surface-enhanced Raman spectroscopy)*

These examples allow emphasizing several complications when attempting to compare systems based on published or manufacturer-supplied accounts of the, e.g., LDC and DR of individual systems. First, if different antibodies are used, then comparisons may reflect the antibody's contribution to the LDC rather than the immunoassay system. Second, even if the antibody is the same, the measurement results will usually contain a component reflecting the skill of the practitioner, and the comparison may represent differences in the experimental skill. Third, different practitioners may use different non-equivalent methods or assumptions in calculating the LDC and DR, resulting in invalid comparisons.

To conclude, the golden path toward getting the right biosensor for a given application still requires access to platforms gathering various types of biosensing systems in order to perform the targeted bioassay (if possible) by the same practitioner using similar biological models to interpret data.

Finally, could the fact that having such a tremendous choice of systems sometimes complicate the choice itself? This is another question that remains to be answered, but we can surely state that MEMS technology will play an important role in the provision of new classes of biosensors for specific applications requiring highly sensitive, portable (thus integrated and low-power consumption) and label-free sensing platforms.

5.1. Bibliography

[BLA 00] BLACK M.H., GIAI M., PONZONE R., *et al.*, "Serum total and free prostate-specific antigen for breast cancer diagnosis in women", *Clinical Cancer Research*, vol. 6, pp. 467–473, 2000.

[GLA 07] GLASS T.R., OHMURA N., SAIKI H., "Least detectable concentration and dynamic range of three immunoassay systems using the same antibody", *Analytical Chemistry*, vol. 79, pp. 1954–1960, 2007.

[HEA 07] HEALY D.A., HAYES C.J., LEONARD P., *et al.*, "Biosensor developments: application to prostate-specific antigen detection", *Trends in Biotechnology*, vol. 25, pp. 125–131, 2007.

[KIM 07] KIM A., AH C.S., YU H.Y., *et al.*, "Ultrasensitive, label-free, and real-time immunodetection using silicon field-effect transistors", *Applied Physics Letters*, vol. 91, pp. 103901, 2007.

[OMI 13] OMIDI M., MLALAKOUTIAN M.L., CHOOLAEI M., *et al.*, "A label-free detection of biomolecules using micromechanical biosensors", *Chinese Physics Letters*, vol. 30, pp. 068701, 2013.

[O'SH 04] O'SHAUGHNESSY T.J., GRAY S.A., PANCRAZIO J.J., "Cultured neuronal networks as environmental biosensors", *Journal of Applied Toxicology*, vol. 24, pp. 379–385, 2004.

[WU 01] WU G.H., DATAR R.H., HANSEN K.M., *et al.*, "Bioassay of prostate-specific antigen (PSA) using microcantilevers", *Nature Biotechnology*, vol. 19, pp. 856–860, 2001.

[ZHA 07] ZHANG B., ZHANG X., YAN H.H., *et al.*, "A novel multi-array immunoassay device for tumor markers based on insert-plug model of piezoelectric immunosensor", *Biosensors Bioelectronics*, vol. 23, pp. 19–25, 2007.

Index

M

N, P, Q

R, S, T